# The Farming Game

*The Farming Game* is the agricultural management text for the 21st century. It provides a sophisticated, pragmatic and comprehensive analysis that is designed to give managers the right focus and clear vision necessary to operate a successful agricultural business.

Taking an integrated approach, this revised edition covers all the vital aspects of farm management including finance, investment, decision analysis, economic thinking, growth, risk and marketing. All of the critical management issues are covered.

This is essential reading for those seeking to manage a modern agricultural business and anyone involved in the agricultural supply chain that needs to understand the importance of farms as the core of agribusiness systems.

**Bill Malcolm** is an Associate Professor at the Faculty of Food and Land Resources, University of Melbourne.

**Jack Makeham** pioneered farm management economics in Australia, and published and taught extensively in farm management economics in Australia and internationally until his death in 1996.

**Vic Wright** has recently retired as Associate Professor from the University of New England and is now a consultant.

# The Farming Game

## Agricultural Management and Marketing

Second edition

Bill Malcolm, Jack Makeham and Vic Wright

CAMBRIDGE UNIVERSITY PRESS
Cambridge, New York, Melbourne, Madrid, Cape Town,
Singapore, São Paulo, Delhi, Mexico City

Cambridge University Press
477 Williamstown Road, Port Melbourne, VIC 3207, Australia

Published in the United States of America by Cambridge University Press, New York

www.cambridge.org
Information on this title: www.cambridge.org/9780521537551



First edition 1993
Second edition 2005

*A catalogue record for this publication is available from the British Library*

*National Library of Australia Cataloguing in Publication Data*

Malcolm, L. R.
Farming and agribusiness : agricultural management and marketing.
Bibliography.
Includes index.
For tertiary students.
ISBN 0 521 53755 X.
ISBN-13 978-0-521-53755-1 paperback
ISBN-10 0-521-53755-X paperback
1. Farm management – Australia. I. Makeham, J. P. (John Patrick). II. Wright, Vic. III. Title.
630.68

ISBN 978-0-521-53755-1 Paperback

# Contents

# Preface

This book is about the business of farmers producing and marketing agricultural products.

The content of this book represents another step in an evolutionary process spanning four decades. In 1971 the late Jack Makeham wrote his original textbook *Farm Management Economics*. In 1981, Jack Makeham and Bill Malcolm wrote the original *The Farming Game*. This was followed in 1993 by *The Farming Game Now*. In 2005, Vic Wright has teamed up with Bill Malcolm (and Jack Makeham) to produce this 21st-century version of *The Farming Game*.

There has been a central and constant philosophy in all of these books, with emphases changing as the world has changed and the views and understandings of the authors have evolved and grown. The central, constant philosophy underpinning all of these works is that economic ways of thinking are at the heart of the interdisciplinary activity known as farm management analysis, and are central to understanding the processes involved in the risky caper of managing farm businesses. Ironically, having economics as the core discipline only works for farm management analysis if practitioners emphasise first the non-economic – the human and technical – parts of farm management. Starting with the farm family and mastering the technology is the pathway to sound farm management economic analysis, and the foundations of the approach taken in this book.

Over time, the farming game has changed, and so have emphases in content of *The Farming Game*. For instance, in the 1993 book *The Farming Game Now*, forward-looking financial management was a significant, enhanced emphasis, moving from towards the back of the earlier editions to the front of the book. In this 21st-century version, *The Farming Game: Agricultural Management and Marketing*, there are two significant changes in emphasis. Most obviously, a new author, Vic Wright, has been recruited to add expertise and insight about food and fibre marketing to the traditional emphasis of Makeham and Malcolm on farm management as a food and fibre production activity. Regarding agricultural marketing, it is emphasised that marketing starts with the decisions about what to produce and how to do it. The nature of agriculture and of farming presents a serious challenge to farmers identifying who are their genuine customers and meeting their requirements well.

The main overall change in emphasis in this latest version of our farm management text is the increased emphasis on investment in innovation and on the business of managing risk – risk and control – to achieve growth in wealth from production and marketing of agricultural products. While investment, risk and growth have always been important in our prior treatments of farm management economics, in this book they are the starting points and constantly recurring themes. In large part, this emphasis stems from our observations of the modern farming world and also from our appreciation of John Dillon's definition of farm management: *Farm management is the process by which resources and situations are manipulated by farm managers in trying, with less than full information, to achieve their goals* (Dillon 1980).

In the prologue to the original *The Farming Game* (1981), Jack Makeham defined the nature of farming when he wrote:

> We didn't call our book 'The Farming Business' or 'The Economic, Technical and Management Aspects of Conducting an Australian Farm Business in the 1980s'. The phrase 'farming game' is far more apt.
>
> 'Game' can mean:
>
> - any arrangement or contest intended to furnish sport, to test skill or strength, or simply to try chance
> - measures planned; schemes pursued; projects organised
> - having an undaunted spirit; unwilling to admit defeat; full of pluck
> - to be happy; to rejoice; to receive pleasure.

As ever, the spirit of the farming game is the spirit of *The Farming Game: Agricultural Management and Marketing*.

# Acknowledgments

Many people have helped us to write this book: professional colleagues, farmers and business people working in farm-related businesses, and friends. They all have played an invaluable role by exposing our thinking to their thinking. Thanks.

As ever, Nanette Esparon has made it practically possible for ideas to be transformed into written explanation and elaboration, and also contributed valuable research. Thanks, Nanny.

Mike Carroll of the National Australia Bank allowed us to use some analysis of returns to agricultural and non-agricultural businesses. Thanks, Mike.

# *Figures and tables*

## FIGURES

## TABLES

# 1 Introduction

Farming in Australia is a fascinating and extremely competitive business activity that abounds in challenges and opportunities; for the best performers, it is financially rewarding and personally fulfilling. The farming game is well-loved most of the time by most of the people in it. Farming is the business of organising and combining and reorganising people and natural resources such as land, sunlight, rainfall and irrigation water, plants and animals with feed, fuel, fertiliser, chemicals, electricity, labour, management skill, specialist knowledge, capital equipment, financial capital and time, in order to produce agricultural products. This is done in ways that more often than not, and for many years, are profitable, are compatible with the investor's view of the risk involved, and help to fulfil the goals of the farmers – all achieved in the face of much uncertainty and risk.

The essential nature of farming has changed little over thousands of years, but nowadays farming *as a business* changes continuously and dramatically in the highly developed economies. In Australia, for instance, farming as a way of using land, labour and capital is increasingly falling into several distinct categories. In one category, farming is a large business activity, usually family-owned, involving tens of millions of dollars of capital investment and millions of dollars of gross income, and is able to earn rates of return on investment comparable to other large businesses in the economy. In another category are smaller, but still substantial, family businesses that also involve quite a few millions of dollars of investment and turnover, and earn competitive returns on capital. A characteristic of the owners of farm businesses in these above-mentioned two categories is that the owner-managers and managers regard themselves more as working 'on' the business rather than only working 'in' the business. Farms of the size described above make up 20–30% of the total number of farms and contribute 70–80% of the annual total value of agricultural output.

The next couple of categories of farms are those that are typical small to medium-sized family-owned enterprises. These farms make up 70–80% of all farms and contribute 20–30% of the annual total value of agricultural output. The medium-sized operations have total capital invested of $2–4 million and earn returns on investment of 3–6% p.a., while the smaller operations face a struggle to earn enough profit to expand and stay in business.

The final category of farming operations is farms owned by people who derive all or most of their income from doing something other than farming, and own farmland for reasons to do with enjoying rural lifestyles. In areas of the country within reasonable proximity of major population centres and attractive natural environmental features, this category of owners of farmland is growing rapidly.

The content of the ensuing chapters of this book is relevant mostly to the situations faced by owners and managers of, and workers in, farm businesses in the first two categories (large and growing medium-sized operations), and to those operators of medium-sized operations who aspire to growth.

While emphasising the central importance to success in farming and related businesses of mastering the technology of agriculture, we have not included much information about general agricultural technology in this text. Farmers today have access to vast amounts of current technical information about every aspect of their daily activities. We emphasise the point that mastering the technology is a necessary condition of success, but this isn't sufficient, on its own, to achieve profits and growth. The associated knowledge that is required – about economic ways of thinking, finance and risk – is the subject matter of this book.

In Chapter 2 we explain why economics is the core discipline of farm management analysis, and why the farm management economic approach is the 'whole farm' approach. This is the approach of considering the human, technical, economic, financial, risk and institutional aspects of how a farm system operates, and of changing the system in order to evaluate all the costs and benefits and net benefits involved. Economic ways of thinking about farming questions are explained and demonstrated throughout the text.

In Chapter 3 we explain the use of tools of analysis to evaluate the health and prospects of a farm business.

Chapter 4 is about analysing the prospects for investment in innovations, which is the key to survival, profits and growth of farm businesses. Return on capital and net cash flow and growth in net worth are the criteria; whole farm budgets, partial budgets and discounted cash flow budgets are the tools.

In Chapter 5, on managing risk and uncertainty, techniques for analysing risky decisions are explained and strategies for managing the main risks in farm businesses are explored.

In Chapter 6, on marketing agricultural products, the focus is on ways in which farmers can sensibly identify their place in the agribusiness system(s) of which they are a part. This is used to explain how analysis can enable valid decision making about whether, and how, capital and expertise can be employed to enhance the value of farm output, given the farm's production capabilities.

In each of the chapters the all-pervasive theme is risk and uncertainty, and control. The emphasis throughout the book is on how to decide what to do that is likely to be best, when there is much that isn't known, much that cannot be known, and much that is beyond the control of the farmer.

In the next chapter, the techniques and ways of thinking about the state of health and prospects for growth of a business are explained.

# 2 The business of farming

In this chapter, readers are introduced to the nature of the tasks and challenges of managing agricultural businesses in a modern market economy. Understanding the business of farming involves understanding the 'whole farm' business system, as well as the economic system beyond the farm gate.

## THE 'WHOLE FARM' BUSINESS SYSTEM

### Introduction

The main components of the farm business system are:

- human elements – goals, labour, management, attitudes to risk and uncertainty;
- technical elements
- economic, financial, growth and investment aspects; and
- risks and uncertainties of the farm system.

The main components of the economic system beyond the farm gate that impact on farm business systems are:

- the behaviour of people and firms in competing businesses;
- suppliers and customers;
- non-agricultural economic sectors; and
- institutional, political and social forces.

In this environment, farmers and their counterparts in the agricultural and input supply chains have to ask themselves, and answer, key economic questions about their businesses. Farm management economic questions a farmer or farm family will want to answer about their business will include:

- What is likely to be the return on all the capital invested in the business, as it currently operates? This is also known as the efficiency, or productivity, of the resources invested in the business.
- What is likely to be the return on our own capital invested in the business, as it currently operates?
- How much is our net worth likely to grow?
- How might we improve the most likely future return on the capital invested in the business?
- Of the alternative means of improving the productivity of the resources in the business, which means are likely to be best?
- What combination of our own and other people's capital is the business likely to be able to service?
- What combination of our own and other people's capital is likely to enable our own capital (net worth or equity) to grow at a satisfactory rate?
- What will be the best means of acquiring the services of land, and what should we pay?
- What will be the best means of acquiring the services of a particular piece of machinery or equipment?
- Should we add capital to the existing land resources that are under our control in order to improve productivity of the whole farm resources?
- How will we set the business up to cope with the reality that yields and quality of product will fluctuate considerably from year to year because of climatic variability, and prices will fluctuate considerably because of market volatility?

The main challenge facing managers of farm businesses is to successfully incorporate new ideas into how they run their business and to be sufficiently flexible, mentally and financially, to continually change how they think about their business and what they do in it. Ability to reorganise in the face of change is the key to survival and success in farm management.

## Farm management analysis

Farm management analysis is a type of intellectual enquiry into changes in resource use on farms. It is a structured process of organising and manipulating information about resources used in farm systems. This information is used to generate further information about the expected extra costs and benefits that are likely to result when a change is made to the way the farm system operates. The expected net benefit of using resources in a farm system in a particular way is compared with the expected net benefit of using the resources in an alternative way. In essence, farm management analysis is farm benefit-cost analysis.

Farm management analyses are carried out in the following segments of the economy:

- *Within farm businesses*: by farmers making what they can of their situation in which much is unknown and unable to be known, where great uncertainty prevails, and where much is uncontrollable.
- *Within public research and development (R&D) organisations*: by people working in R&D in the fields of science, agricultural and natural resource science, agricultural economics or rural social science.
- *Within private rural input supply and output processing businesses*: by researchers and providers of goods and services representing both established and new technology used in farm production, and by firms adding services to farm production.
- *At sources of information to farmers*: by publicly funded and private business people who provide information directly to farmers as advisers, consultants and providers of education services. They operate professionally in between the farmers and those primarily involved in farm-related R&D.

There is an overlap of people and flows of farm management information – and misinformation – between these arbitrarily defined segments of the rural economy involved with the analysis of choices relevant ultimately to the management of farms. Usually, in investigations of farm management questions, great effort is made to ensure 'good science'. However, 'good economics' is just as important.

When done in accordance with the tenets of appropriate theory, the information generated by farm management analysis informs the decisions of managers of farm resources in ways that are likely to contribute to them achieving some of their goals. The alternative approach – analyses that violate tenets of economic theory – are likely to generate information that leads to conclusions, decisions and actions that do little or nothing to advance the cause of researchers and farmers achieving their goals. Too often, analyses of farm management questions show little evidence of knowing, first, that the maximum isn't the optimum, and so the science/technical emphasis on maximising physical output per unit input is flawed; second, that the future is a different world, and so the accounting focus on looking backwards at averages and minimum average costs of production is also flawed; and third, that the whole of the farm system is the domain on which to focus when analysing changes to farm management practice.

In the conduct of farm management analysis, the textbook representation of economically rational decision-making behaviour of managers is but part of the story. In practice, usually and sensibly, decision makers also draw on other sorts of knowing. Constraints of time and resources and 'ability to know' dictate that a 'fast and frugal' (Gigerenzer and Kerr 1999) approach to decision making has to apply. Furthermore, of course, a bad decision can turn out to be the right decision through the intervention of chance, and vice versa. In an uncertain world, relatively simple analysis based on a few key bits of information is the practical way to go – but the economic logic has to be right! The case is made throughout this text for good decisions based on sound analysis of the important relevant information obtainable by the decision maker, given the constraints of resources and time. Sound approaches to decisions will contribute to decision makers achieving more of their goals than will the alternative approaches of acting randomly, or worse, conducting consistently bad decision analysis and hoping to be consistently lucky.

## The whole farm approach

Analyses of questions about choices, or problems, cannot be useful if the focus is on only the technical parts of the question and other parts of the question that are significant, such as economic forces or social conditions, are ignored. The philosophy of farm management economic analysis is that it is more useful to identify and solve the whole of a farm business problem in an approximate manner, than it is to identify and solve a small part of a problem extremely well while leaving out significant parts of the question at hand. For instance, it is of little value to identify that the grazing management of the livestock system is poor, or that the state of the soils or pastures isn't state-of-the-art, and therefore must be the main problem. It could be that the main problem is that the equity of the owners in the business is so low that the annual debt servicing requirements exceed the annual cash surplus expected to be available for servicing debt; or the farm family may be dysfunctional with conflicting goals; or the farm managers may be at a stage of their career when they are no longer motivated to pursue innovation or growth. (They may already know how to farm better than they do!)

Farm management analysis is an interdisciplinary activity: the human element is included and is the starting point; the technical basis of agricultural economic analysis has to be sound; economics is the integrating and core discipline; the business has to be financially feasible; risk and control in the face of uncertainty and volatility permeates all activity; and the role and influence of economic, institutional, political and social forces beyond the farm have to be recognised fully. Knowledge from, and an emphasis on, all these facets of agriculture culminates in the ability of managers to reorganise farm resources and to succeed.

In the process of explaining how firms operate in the economy, the aim of agricultural economic analysis is to put the components together into a whole, albeit a simplified whole, but one in which all the elements that have an important bearing on the question are considered. In contrast, science often focuses on part of this whole but, while dealing completely with and 'fixing' the small part of the whole, has difficulty explaining how the whole business, or the relevant and connected parts of the agricultural economy, works. Narrowly focused science has limited capacity in prescribing solutions to problems of the whole business because there are no technical solutions to farm problems, only technical components of people-based solutions to what ultimately are 'people problems'. The farm management economic approach is thus the whole farm approach. In practice, there is no other approach to identifying correctly, and solving, problems of farm business systems. The method is to start with the people; sort out the technical system; and then analyse the benefits and costs and finance and investment and growth and marketing prospects of the whole farm system in light of the important risks and uncertainties and the major influences that are beyond the farm gate and so beyond the control of the farmer.

## Economics: The core discipline of farm management analysis*

The key task of farm management is making choices between alternatives. Farm management analysis is about analysing those choices. Economics is the discipline of choice. Economics entered farm management analysis from the middle of last century, and became the core discipline of academic work in farm management. In the context of farm management analysis, what does 'core discipline' mean? It is the discipline that organises the practically obtainable relevant information about a question, or series of questions, into a framework and form that enables a choice to be made. The choice is between alternative actions faced by management that, in the light of the goals, is informed, reasoned and rational.

The ways modern market economies operate now reflect the influence of insights of great economic thinkers over several centuries. Less prevalent, and less influential, is the economic way of thinking about farming choices that has been rigorously developed since the mid-20th century by some major thinkers about farm management economics. The relationship between economics and applied farm management analysis has been neither comprehensive nor consistent over time.

Economics encompasses a number of key sub-disciplinary areas that are particularly significant for the management of farms. These disciplinary areas are farm production economics (input–output relationships), risk, finance, marketing, time, and the microeconomics of choices and actions of groups of firms responding to market forces. Farm management analysis encompasses considering alternative actions under risky and uncertain circumstances. Economics – the discipline of choice – is central to farm management analysis (McConnell and Dillon 1997). Choosing between alternative uses of resources draws on a number of key economic principles – namely, comparative advantage, diminishing marginal returns, equi-marginal returns, cost analysis, opportunity costs, input and output relationships, size and scale, gearing and growth, risk, time and tradeoffs between goals. Economics is needed to bring the many relationships of a system, and between systems, to some common unit or basis of comparison. If this isn't done, it isn't possible to analyse systems meaningfully or to compare alternatives meaningfully in terms of expected benefits and expected costs. That is the first reason economics is the core discipline of farm management analysis.

* Some of the central ideas in this book are dealt with in B. Malcolm, 'Where's the economics? The core discipline of farm management analysis has gone missing!', Presidential Address, 48th Annual Conference of the Australian Agricultural and Resource Economics Society, 2004, in *Australian Journal of Agricultural and Resource Economics*, September 2004.

Farm systems are dynamic and complex. The second reason economics is the core discipline derives from the rigorous, abstract and conceptual nature of economic enquiry. The emphases in economics on the counter-factual and the counter-intuitive, and on the subsequent rounds of cause and effects, go a long way in helping to clarify understanding of complex, dynamic, whole farm systems. Economic principles tell what information is needed, and conveniently organise such information in ways that suit analysis of benefits and costs. Most importantly, the logic of economics helps in defining the question in a way that facilitates finding solutions. The question is the answer!

The third reason economics is the core discipline of farm management analysis is that economics sets much of the agenda for the decisions that have to be made. Knowledge and techniques from economics are combined with empirical data to help make decisions about *what* to produce and market, the *method* to use in producing and marketing, and *when* to produce and market farm product.

Finally, the main focus of farm management is the implementation of new ideas and production technology amid reorganisation of the farm business in the face of powerful and continual market forces for structural change. Factors beyond the farm gate, in markets, play a bigger role over time in determining the extent to which farmers' goals, such as wealth accumulation, consumption and leisure, are achieved over time. Indeed, such forces can be as influential as the actions farmers take within their farm boundaries. Components of the larger economic picture, including changing the comparative advantage of competitors, the cost-price squeeze, and pressures for adjustment and adoption of new technology, are critical to farm management analysis and farm business success. All of this happens in an activity with limited scope for product differentiation, such that the conventional tenets of business marketing commonly do not apply.

The discipline of economics plausibly explains the setting for and influences on behaviour of many agents (producers/firms and consumers) beyond the farm gate. Economics facilitates plausible conjecture and expectations about the behaviour of competing and complementary businesses, and about changes in industry structure. It anticipates to a degree the external forces for internal change on farms. Keen appreciation of wider economy phenomena and forces brings valuable insights to decisions about opportunities created for farm entrepreneurs by counter-cyclical behaviour; to asset valuation; to financing, gearing and growth decisions; to activity mix choices; to investment timing; to intensification and extensification; to risk diversification; and, of course, to the increasingly important off-farm investment portfolio decisions. Thus the fourth reason why economics is the core discipline of farm management analysis is that in economics, the external effects of markets, time, growth and dynamics on the internal working of farm businesses are confronted explicitly.

Making the case for economics being the core discipline of farm management isn't a case of disciplinary imperialism; nor should it be seen as implying a narrow, unbalanced approach to farm management. The 1987 Nobel Prize winner in economics, Robert Solow, explained the strengths – and limits – of economic analysis as follows:

> The true functions of economic science are best described informally, to organise our necessarily incomplete perceptions about the economy, to see connections that the untutored eye would miss, to tell plausible – sometimes even convincing – causal stories with the help of a few central principles, and to make rough qualitative judgements about the consequences of policy and other exogenous events . . . the end product of economic analysis is likely to be a collection of models contingent on society's circumstances and not a single monolithic model for all seasons. (Cited in Fitzgerald 1990, p. 21)

Substitute 'farm actions and goals' for 'policy', 'farm' for 'economy', and 'farm family' for 'society' and what Solow says about economics applies equally to economic analysis at the level of unique farm systems. In the context of farm management analysis, it just so happens that at the level of sensible analysis of farm choices, the key theoretical principles to do with marginality, costs, time, investment and risk are well established, and estimates of key economic parameters can be made. Theory about equally important but less congenial elements of farm management analysis, such as uncertainty and non-material goals, still has quite a way to travel. Still, an important aim in the rest of this book – similar to how Solow says it – is 'to tell plausible – sometimes even convincing – causal stories with the help of a few central principles, and to make rough qualitative judgements about the consequences of risky farm management and marketing decisions'.

## The human conditions: Goals, labour and management

### *Goals*

When considering why members of farm families do or do not take certain actions, the starting points are with their needs, wants, stage of life, history, goals, and views about risk. The goals of members of farm families ultimately determine how properties are managed and how the farm families might exploit the potential of farms. Knowing the goals of both the farmer and the family helps to explain why the farm is managed in the way it is. Goals also help to indicate how (if at all) the farmer might be able to exploit the potential of the farm.

Common goals of farmers are for the business to survive and to grow, and to set and to overcome challenges. Other goals are numerous: to farm well and to be recognised for this; to improve the physical state and appearance of the farm; to acquire extra land or to control a larger business for the future and for heirs; to have a reasonable but not profligate standard of living which compares reasonably with others in farming and society at large; to earn enough profit to be able to improve and develop the farm so as not to have to work so hard as they age; to achieve capital gain and increase wealth; to help to protect, preserve and improve the wider natural environment in which they live and work; to have good-quality animals and crops in good condition; to reduce income tax; to have a satisfying rural way of life; to have children well educated (often, better educated than themselves so that they have the option of not farming); to have enough leisure, increasing over time; to be a respected member of the community; and to have enough money to pursue non-farm interests. Importantly, some of these goals are in conflict; thus tradeoffs are involved between achieving various goals at various levels.

A role of professional farm management advisers is as 'professional goal adjusters'. Farmers might state definite objectives but might not see that their physical and managerial resources fail to match their wishes. It is essential to determine if the product markets and the resources of land, technology, capital, credit and skills are compatible with goals. If they are not, then the goals might have to be modified or re-defined. Sometimes, farmers might not be aware of the possibilities for exploiting the full potential of the resources that they and their families operate. They might have 'vistas unperceived' and may have to be persuaded to raise their expectations. Others might be content and not have high hopes. Advisers can have an inspirational role when the goals are limited compared to the potential of the resources under the farmers' control. Mostly some kind of adjustment about stated goals has to be made when the situation is assessed, taking full account of the agricultural, economic, financial, risk, growth and human realities. Given this, however, the adviser cannot know more than the farmers do about what is 'good' for them and how best to reorganise their affairs.

Most farmers are moderately profit-oriented, and take a medium-term view of using and husbanding their resources to the extent that more immediate economic forces allow. A reasonable supposition is that most farmers are largely motivated to do a bit better and to make a bit more gain in the short and the long run, when most, but not all, of that gain is profit. This applies as long as they don't have to sacrifice too many of the other things they value highly, such as health, family life, leisure and outside interests. 'Farm to make a living and you can get a good living; farm to make lots of money and it will often deny you a living' and 'farm like you are going to live forever' are age-old adages whose truth is widely recognised in farming communities.

### *Labour*

The numbers, ages and skills of farm families and workers have to be considered in analyses of farms. Of particular interest are the methods used to meet peak workloads and the skills available to do such specialised tasks as basic maintenance, repairs and adjustments to machines, pumps and engines, or the care of animals and crops.

Identifying where the strengths, weaknesses and preferences of the labour force lie in performing the various physical and mental tasks helps to explain the way farm businesses are being operated and how well these are performing. The skills and judgments of the operator are among the main inputs to farm production. These mostly determine the potential profitability of the businesses. Importantly, if changes are being considered, it is vital that the farmers have the necessary skills, knowledge and personal make-up to handle the changed, often more intensive and technically complex, situation.

The extent to which casual and contract services are used is important in determining the output per permanent labour unit. The way in which workloads and specialised tasks are met form part of any farm analysis. In the appraisal of the potential of the farm, determine whether the existing permanent labour force, as well as convenient and reliable contract services and casual labour, can supply the skills needed to handle an intensified or more diversified operation. If not, ask what training is needed to equip them to cope with the new situation. Would additional skilled people have to be hired? Possibly, the numbers and skills of the farm family labour force could be used more productively if the farm were diversified more. There may also be the potential for engaging in some non-farming activities.

### *Management*

The principles of management discussed below apply to any business, though in this case the principles are applied to farm and agricultural-related businesses. Sound business management analysis and decision making require an approach where all the important aspects of a situation are considered. This is called the interdisciplinary approach. To understand how a business 'works', it is necessary to take full account of the combination of land, labour and capital, the goals and interests of the owners, the available and obtainable skills and resources, as well as the many risks and uncertainties involved in the business. Every business is unique. At any time, the mix of resources available to any business – the people, land, climate, credit, animals, location – is unique. Importantly, so is their history; where a farm business has been in part determines the shape of things to come.

People in charge of businesses have to decide what to produce and how to produce it; what investment projects to make; and how best to manage the risks and uncertainties. Decision making remains the main task of those running a firm – and decisions are taken in the face of considerable risk and uncertainty. The management of systems involves identifying constraints to goals being achieved; identifying and evaluating alternative strategies to achieve goals; implementing the selected strategy; monitoring technical or financial performance; comparison of the expected outcomes with the actual outcomes; and responding as the future turns out differently from what was hoped and expected.

The management process can be summarised as attempting to deal with uncertainty and risk by: (a) planning and deciding; (b) organising resources and doing the things that have been decided; and (c) keeping a close watch on how things are going and responding to the changes that continually arise. This is called control, and is possibly the main function of management because management is essentially about recognising and adapting to changes in circumstances, often before the changes have occurred, and when the full extent of the changes that are happening are not known.

Information problems bedevil these processes of management. In trying to understand how people who are running farms behave, it is important to recognise the inadequacies of information with which management has to manage. Sometimes, after an event, it may not be clear whether the correct decision was made or whether a decision was made for the right reason. Forecasting is one unavoidable element of management decision making, because the ultimate effect of management decisions invariably depends on the effects of factors to do with the future that cannot be known with certainty, if at all. Forecasters rely a lot on what happened in the past, but their understanding of a situation in the past is never complete, and neither is their understanding of the current situation. It is never totally clear which facts today are relevant to guessing what might happen in the future. At best we have a good knowledge of the fundamental forces and key principles that explain linkages between influential economic factors.

Further, uncertainty means that managers can never know what will be the full consequences of decisions. There are uncertainties about technology and about outcomes. When problems are complex and the costs of getting it wrong are large, it is useful to use structured decision processes. This involves identifying the main sources of risk and uncertainty and the main alternative actions. It involves identifying and evaluating a range of possible consequences and, on the basis of these evaluations, making some judgments about what is most likely to be the best thing to do.

**Labour management**

A key part of management is labour management. Owners need to direct and motivate staff with appropriate incentives for managers to act on the owner's behalf in the interests of the business and the owners. This is also called the principal–agent challenge. The principal, who is the owner, has to motivate the agent, who is the manager, to act in the interest of the principal. Complicating this is the reality that the attitude to risk and uncertainty of the manager will differ from the attitude to risk and uncertainty of the owner. And, managers may have a different view from other members of the workforce about certain matters. Therefore, directions and incentives have to be established and communicated well to all, so as to enable the owner's goals to be achieved through the performance of the management and the workforce.

Communication and reward, respect and trust are the bases on which successful labour–management–ownership relationships are built. Without good communication and mutual understanding between the owners, managers and staff, the best of plans won't work. Within the family workforce, good communication is also critical. However, the means of attaining good communications where it doesn't exist is usually difficult because of complex and close interpersonal relations. Much of the skill of getting things done by people (communication and leadership) depends on personality, fairness and respect-earning behaviour by management. Most of these traits are innate. The role of innovative ideas in motivating, rewarding and retaining good-quality labour is growing. Employee remuneration structures are being developed that provide high-quality staff with the incentives to perform at a high standard and to stay with the business – for instance, setting up a trust into which a portion of employee bonuses is paid and which then acts as a 'golden handcuff' to help retain the employee. Ultimately, it is the personal qualities of the individuals involved in conducting the business that determine the results. The vital aspect of management is the task of getting the most from each person according to their ability and rewarding them appropriately.

**Good management**

Many skills are important in good management. The most crucial skill of farm management analysis and decision making is identifying correctly the true nature of the problem. Pertinent information, processed by sound analytical and planning techniques, makes for good problem identification and good decision making. Increasingly useful information is becoming more available, more quickly, to more farmers, than in previous decades. The outstanding feature of the best farm managers is their mastery of information about the whole spectrum of the process called farm management. Of the skills necessary in farm management, how much can be taught and learned? Technical skills, both applied and theoretical, can be learned. 'Economic' ways of thinking, the ability to recognise the essence of a problem and to draw on general principles and budgeting techniques, can be learned. So, too, can the ability to communicate clearly. The human side of management, the personality, understanding, intuition, ability to see an issue through another's eyes, and other intrinsic human qualities, are probably difficult to learn or to radically change other than through experience.

Management skill, business acumen, entrepreneurial skills, intelligence, shrewdness, judgment, mastery of information, ability to assess and cope with riskiness, and intuition are all vital features of top managers and are mostly innate qualities. Individuals might or might not have them, and they probably are not gained directly from formal teaching and learning, although the disciplined ways of thinking that comes from study can be very helpful. But many analytical skills can be demonstrated and applied in formal teaching and learning situations to help owners of farm businesses avoid some of the basic errors that are made regularly by so-called entrepreneurs and managers in Australia's non-farm businesses and financial institutions. One feature of managing farms that distinguishes it from managing many non-farm businesses is the close relationship between the household (consumption unit) and the business (production unit). This means that non-production issues have a larger role in how the business is managed than is the case with many non-farm businesses.

Some characteristics of good farm managers are that they:

- are aware of all the relevant information about whatever particular tasks or projects are at hand;
- know the technology thoroughly and keep up with new technological developments;
- emphasise getting jobs done on time;
- think ahead (using a bit of paperwork helps here);

- discuss ideas, procedures and alternatives with others, and make provision to cope with the unexpected;
- have a system for keeping up to date with daily work achieved, for regular consultation, and for ensuring that employees have regular, clear instructions;
- have vision;
- are adventurous, but sound;
- can control costs;
- have sufficient grasp of farm management to decide on well-analysed and economically sound plans of action; and
- can carry out plans.

**The rules of the game**

For any manager, the beginning of rational analysis of the choices facing them is understanding the determinants of performance of a business; what it takes to achieve the objective. In broad terms the determinants of business performance are how appropriate the behaviour of the firm is to the relevant part of the environment in which the firm operates. This includes two related domains: the choice of what to produce, and the efficiency with which it is produced.

There is an unavoidable sequence, or hierarchy, of choices here. The choice of output implies a commitment to create and maintain capabilities of the firm of specific relevance to that output. When a farmer decides to produce fine wool, for example, requirements for physical capital, technical systems, human skills and marketing systems are all implied by that decision. Therefore, the decision about output creates impediments to later change in output. This decision is intrinsically 'strategic'. It is inevitably long-lived, because changes to it are slow to put into effect, and it places limits on farm performance. If the decision turns out to have been a bad move over a meaningful period, no amount of cleverness in managing the farm can cause performance of the business to exceed the limits imposed by the bad choice.

Good farm management, like the good management of any firm's productive effort, can only deliver what choice of output defines as being the maximum potential returns as the future unfolds. Bad farm management, though, can lead to much waste within that boundary and can even cause wise choice of output to lead to very poor performance.

Competitive forces drive all firms to ensure that, in the long run, they create output that is competitively rational. Output must be demanded and must be capable of being produced with a competence that ensures sufficient profit to sustain the firm. When we note the relevance of non-financial goals being important to farmers, and other managers, it is important to realise that financial performance has to be sufficient to allow the firm to persist.

The variability that characterises Australian agriculture bedevils rational decision making at both levels: what should we produce; and how should we produce it? The reason is that the control over performance available to the manager is weakened by variability. To be more precise, it is *unpredictable* variability that is the problem.

Variation is just variation. When we can predict with confidence what variation we will encounter, there is no management difficulty. When we know that there will be variability, and of what scale, but we don't know when or how often, we have a problem. (Drought is an example.) When we don't even know the scale of the variability, we have a *real* problem.

The uncertainty and consequent variability in financial performance facing farms means it doesn't make a lot of sense to be highly sensitive to short-term changes in the operating environment. It is necessary to watch for trend changes that might occur in relative prices or production risk, but successful farm management involves working to a multi-period commitment to activity which, in the medium term, is expected to deliver the desired financial outcomes.

We consider managing risk and uncertainty later. For now, we can note one implication of the threat to performance of the business that variability creates: the costs of failing to understand the knowable and predictable aspects of the farm and its operation, and of its sensitivity to variable factors, are potentially very high. The costs of avoidable management clumsiness are amplified by variability in the uncontrollable aspects of the management environment.

## Technical systems

This section outlines the approach to analysing the way the technical aspects of the animal and crop production systems that make up farm systems operate and the type of information required. The approach is to look at the present situation and then identify the potential, and the constraints on achieving the potential, of the farm system.

### *Animal production systems*

In this section is examined the present and potential of animal production systems' activities, and the constraints on those activities.

**Present feed conversion activities**

In analysing feed conversion activities, determine what system the manager uses to convert feed into saleable product and income. On most farms this is done through animals, but some enterprises convert it directly into such saleable products as hay or pellets. The potential for these activities seems to be increasing, as animal industries become more intensive. It is worth looking at the scope for these forms of feed conversion.

Consider the chief features of animals as feed conversion mechanisms:

- type of animal activities;
- structure of the flock or herd, by age, sex and numbers; and
- system used to replace the herd or flock.

There are two systems of replacement of animals:

- breed own replacements; and
- buy-in replacements.

There are three systems of feeding:

- grazing (set stocking or some form of grazing involving regular movement of animals on to and off pasture);
- hay, silage, fodder crop, grain, mixed rations, agistment; and
- totally or partially fed on purchased concentrate feed.

The husbandry of animals is critical to efficient conversion of feed to saleable output. Husbandry covers the use of medications and chemicals; veterinary services; the method of harvesting the product (milk, wool, eggs); the supervision and management at mating and birth; and the time and interval (frequency) of parturition (calving, lambing and farrowing). As well as proper husbandry of livestock, increasing the genetic potential of animals, and then providing the feed to fulfil that potential, are the keys to increasing the output of saleable product. The use of genetics includes artificial insemination (AI), embryo transfer selection based on performance recording, use of progeny-tested sires, cross breeding, and purchase of special hybrid animals, as in the case of the intensive industries.

**Potential feed conversion activities**

In the analysis of the animal-based feed conversion system, consider the potential and the scope for improvement. The opportunity to improve performance by exploiting new or different systems of feeding is worth investigating, because it is in this area that there are some of the greatest gains in profitability. In animal activities, it is necessary to challenge traditional methods and to consider whether there is a role for fundamental changes to the type of activities, or to the way existing activities are operated, to tackle biological and profitability problems such as poor reproduction or growth rates or feed costs. With herd or flock structure, there might be potential for increasing the ratio of female breeding stock to males, or for selling or mating animals at a younger age. The replacement system might be improved by changing partly from a self-replacing system to buying hybrid animals of superior performance, or by mating cull females to sires of another breed.

Husbandry practices on many farms can often be improved by using the latest developments in the rapidly changing technology of control of internal and external parasites; in shearing, milking, poultry shed, and piggery design and operation; and in the development of labour-efficient practices of animal husbandry. Medicines, chemicals and veterinary advice are also integral to insect and parasite control, and general animal health, such as meeting minimum requirements for essential nutrients and trace elements. These are often very efficient when the improvement in animal performance and 'profits' are considered. The first need is for animals to be properly fed, as poor nutrition makes them more susceptible to disease and parasites. Capitalising on the various techniques for improving the genetic worth of animals is also usually efficient. Time and effort spent on investigating means of achieving genetic gain gives good returns, as does appropriate identification of animals and the recording and use of information about individual animals.

Using feed supply efficiently is the key factor affecting the profitability of animal enterprises in both grazing and 'factory farm' production. There are numerous possibilities for ensuring efficient use of feed – for instance, timing of lambing or calving, more controlled grazing, segregation of age groups, using feeding facilities that reduce wastage, and mixing different species of grazing animals (cattle, sheep and goats). Strategic feed supplementation is also very useful – for instance, during the last stage of pregnancy, early lactation and weaning. The quantity and quality of dry matter produced and utilised per hectare is the key to profit in grazing systems.

**Constraints on feed conversion activities**

Some important constraints on the exploitation of the potential for improving animal enterprises include climate, environment and risk of disease.

Replacement through bought-in replacement stock, rather than through a self-replacing system, depends on being able to acquire the quality replacement stock of the type that is required. This method can also carry increased risk of introducing disease. Replacement systems have implications for income tax, too, as the tax system favours self-replacing systems.

Barriers to using feed more efficiently include not knowing the best way of managing grazing animals, a shortage of labour or proper facilities (sheds, yards, equipment), and the risk and other changes associated with changing the timing of lambing or calving and operations. Often, better husbandry practices are not possible because of lack of competent people to care for the animals.

The main restraints on exploiting genetics to the full are environmental limits preventing the genetic potential of the animals being expressed fully, or the animals already being adapted to maximum performance in that environment; ignorance of the practical implications of genetic research; prejudice against crossbred animals; and the absence of AI skills and facilities within easy reach. There is also the problem, where introducing new strains (not species) is wanted, of finding those strains that are as well adapted to the local environment as the strains that have been traditionally used.

**Present feed supply**

The two main sources of feed supply are pasture and the concentrate feeds used in intensive industries. In an analysis of a farm pasture supply, look at: the topography (whether steep, undulating or flat); the soils and nutrient status; the presence of rocks and non-edible vegetation such as trees, thorny shrubs and unpalatable bushes, and the incidence of plants toxic to animals. Next to be assessed are the pasture species present – their palatability and digestibility, density, and nutritional value; what crop residues are available, and when; and whether special crops are grown for animal fodder.

Topography is very important. The feed supply varies with the slope and aspect of the land. An almost vertical slope facing south doesn't have the pasture production of a river flat. Even so, a balanced combination of flats and hills can often provide more flexibility if grazing pastures in different seasons.

Soils vary greatly in structure, in nutrient status, and in fertility. Soils can be classified into major and minor groups according to their ability to produce pastures and crops. A breakdown of the area and range of soil types on a farm is essential for physical analysis: this is the basis of both present and potential land use.

**Present soil nutrient status and fertiliser history.** The most common deficiencies in Australian soils are phosphorus and nitrogen. Soil tests indicate the pH (degree of acidity or alkalinity) and the levels of the major elements phosphorus, potassium and sulphur and are fairly reliable for use in a fertiliser program. Tissue tests of green growing plants are more useful tests for the significant trace (or minor) elements such as molybdenum, copper and zinc. The fertiliser history is useful mainly to help detect deficiencies other than phosphorus. If a farm has had a history of a tonne of lime per hectare and 2 tonnes per hectare of superphosphate over the previous five years, and has raised both soil pH and soil Olsen P (measure of available phosphorus), and yet the pastures are not vigorous, it is likely that some element other than phosphorus is lacking and limiting production. In practice, a combination of soil and plant tissue tests every few years, records of past fertiliser use, some simple, moderately sized trial plots, and the farmers' observations of how the paddocks have been responding to recent applications of fertiliser, is an effective way of determining present fertiliser needs. Off-farm implications of fertiliser use are becoming increasingly important, especially in activities with high fertiliser inputs and the chance of nutrient run-off into waterways.

**Pastures.** The following aspects of the pastures have to be defined:

- composition: native, introduced legumes, grasses, weeds; balance between legumes and grasses, and any remnant vegetation considerations;
- density of plant population;
- vigour, persistence of species, expected life;
- suitability for various classes of livestock – for instance, non-breeders, young growing animals, breeders, sheep or cattle;
- weed control: when, how; use of sprays and grazing;
- pasture improvement, establishment methods; and
- scope for adoption of elite genetically superior pasture species specially bred for characteristics such as winter growth or disease resistance, and so on.

Based on rainfall records and farmer experiences, there is for each region an average expected pattern of seasonal pasture production, as a result of the rainfall pattern. Thus, in southern Australia, rain-fed pastures usually begin active growth in late autumn, slow down in the cold of winter, reach a peak in spring, and don't grow in summer and early autumn. The further north the region, the greater the tendency for pasture production to be summer-dominant and winter-dormant. Generally, the lower the annual rainfall the greater the degree, and impact, of unreliability or variability of pasture. The expected average seasonal pasture production has to be defined because this determines the system of animal production that is best suited to using the pastures most profitably. The scope to move the predominant pattern of pasture feed supply is worth investigating. Another often-important aspect of feed supply is whether some of the feed comes from agistment or leasing land; the reliability and tenure of this source of feed needs to be determined.

Rarely does the average season occur. For example, in southern Australia, the opening rains, on average, fall during April. Yet these can begin as early as February or as late as June. Winters in this region can be very wet, with the paddocks becoming waterlogged, which impedes pasture growth and use; or they can be very dry and warm, resulting in good winter pasture supply. There are similar variations about the average season in other areas, often with wider fluctuations.

The strategies and particular tactics that the manager of an individual farm adopts to meet these variations about the average vary greatly, as do their profitability (or loss-reducing) effects. The general aim is to reduce the harmful effects of adverse situations and to exploit favourable circumstances. Thus it is essential, when analysing a farm, to discover what measures the operator adopts to handle variability of feed supply – for instance, supplementary feeding, stock trading, fodder conservation, and reduction in body condition of stock. Skill at managing the fluctuations about the normal pasture supply is one of the main requirements for successfully running pasture-based production enterprises.

**Concentrate feed and complete mixed rations.** The chief issues in regards to concentrate feed and complete mixed rations are whether it is best for the operator to mix the rations or to buy pre-mixed feed from a miller; how much labour is saved from using pre-mixed feed, and the relative quality of the two sources of feed. The trend is now to use pre-mixed feeds, and to use the labour saved to do other things such as run more stock. The person studying the situation on the farm needs to determine if the grain is home-grown or bought; how the best mixtures of grains and additives are determined; and the methods of mixing, storing and feeding. Where feed is bought from a company that mixes feed, the reliability of the feed supply, its quality and the consistency of quality are important. It is necessary to ascertain the relative performance of animals on home-mixed feed and factory-supplied feed, and the relative costs.

### Potential feed supply

Increasing feed quantity and quality by better species and fertiliser use is the single most important action available to farmers who manage pasture-based animal production systems. Much technical information exists to help farmers implement the physical part of a development program if they decide that it is economical to do so.

Winter and summer fodder crops are often valuable as a means of increasing the supply of high-quality feed when seasonal pasture is low. The scope for exploiting the potential contribution of fodder crops has to be examined in any farm analysis. Critical here is assessing the quantity of feed that a crop may provide.

On pasture-based farms there is increasing interest in using supplementary irrigation from water harvested from run-off that would normally be wasted. As well, a number of engineering developments have led to supplementary irrigation schemes that can be run more cheaply than by using traditional methods. Generally, because of its cost, feed grown using supplementary irrigation schemes needs to be fed to animals that can convert it into a high-value product such as milk, stud stock, lamb or veal. It can also be used as hay or pellets as part of the ration of intensive animal systems such as pigs and poultry.

### Constraints on feed supply

The main constraints on getting the full potential feed supply are the costs of seed, fertiliser and growing fodder crops, relative to the amount of extra feed that grows as a result of using such techniques, and the use to which the extra feed is put. If lots of extra feed grows, and it is used efficiently by animals whose product (such as milk or carcase) is bringing a high price in the market, then it could be worth doing. Otherwise, it could be a waste of money. Also, it might not be possible to match feed requirements to feed availability efficiently. Another constraint can be lack of knowledge about livestock and feed prices, particularly in situations where the farmer wishes to exploit the good years and cut losses if the season is bad.

The main barriers to achieving the potential for improvement in grain-based feeding systems are lack of specific knowledge about particular animal production methods and requirements; the absence of readily available feed-testing services; poor information about energy partitioning within the animal; poor knowledge about the energy performance of feeds in combinations; and inadequate attention given to animal health. Animal health is usually a bigger issue in intensive feeding systems than in extensive systems.

### *Cropping system*

The cropping system is analysed in the context of the timing and sequence of crop activities on land areas, and the interactions with related activities such as animals and fodder crops. The present state and potential and constraints are looked at.

#### Present cropping activities

The main points that need to be examined when analysing the cropping enterprise on a farm are:

- the soil types and nutrient status;
- the types of crops grown – whether they are (a) short-term (three- to four-month) crops such as tomatoes, (b) annual crops such as wheat or sorghum, (c) longer-term (18-month) crops such as sugar cane, (d) crops that take three to four years to reach maturity, or (e) tree crops which grow for many years (for example, apples);
- the cropping system that is being followed, the sequences of crops that are grown, and the methods of preparing, planting, husbandry and marketing used for each crop;
- on orchards and in forestry, the age of the trees, and the expected life of each stand or block, have to be defined; and
- in cropping areas, the crops grown also need also to be classified into winter and summer crops.

**Maintaining soil fertility.** Most progressive farm managers aim to have a cropping system that yields consistently well while not reducing the long-term fertility and erosion status of the soil. However, there are many farms where the arable soil has been plundered. It is essential, when examining the cropping activities, to find out what (if any) measures are being taken to maintain soil fertility. Fertility is defined in terms of structure, freedom from weeds, nutrient status, absence of harmful micro-organisms, and erosion. The question of how best to maintain fertility is critical on intensive vegetable farms, as it is also to orchard farmers and growers of plantation crops and broad-area grain crops. The most common method of maintaining fertility on broad-area farms is to have legume–crop sequences.

In some areas where crops are more profitable than stock, there is a trend to have non-legume or reduced-legume systems of maintaining soil fertility. Among the techniques used are stubble retention to maintain structure and reduce erosion, and applying nitrogen fertiliser or using herbicides to control weeds, and chemicals to control harmful micro-organisms. Minimum cultivation techniques using herbicides are increasing in favour. Assessment of the scope for expansion of one or other of these two basic systems to maintain fertility in the future is a vital step in any farm appraisal. On intensive vegetable farms where there is little crop residue to work back into the soil, gypsum and poultry manure might have to be used to help maintain the structure of the soil.

**Irrigation system.** The chief aspects of the present irrigation system that need to be looked at are the amount of water available, water quality, salinity problems, labour requirements, the system of delivering water to the growing crop, and drainage. Of particular interest is the reliability of the water supply and water quality, and the total cost per megalitre of water. Can water be purchased? For annual use or permanent purchase? At what price? Salinity and some idea of the content of toxic materials in the water must be known.

**Potential cropping activities**

When evaluating the potential for the crop enterprise, a number of things need to be examined, including: the scope for new crops; changing the mixture of crops; having better sequences of crops; and intensifying cultivation (say, through shorter rotations and sequences and land improvement). It might be biologically better to change the mixture of crops. Increasing the proportion of summer to winter crops might lead to better weed and disease control.

Having looked at what is being done on the farm about soil fertility – whether it is being 'plundered', maintained or improved – it is then necessary to see what potential there is for improving the situation. It is important to ask these questions:

- Can more, or better, legume crops be grown?
- Can more crop residues and stubbles be returned to the soil, either directly or through grazing animals?
- What happens if more cash crops and fewer legume crops are put into the rotation?
- How can fertility be improved by using more chemical fertilisers and sprays in a better way?
- Could more grazing animals be introduced into the land-use system?
- What role is there for strip cropping or reduced tillage?

As water is usually a scarce resource, the way in which it is distributed between the various alternative uses for it – annual crops, tree crops, animals, households – has to be analysed to find the most effective use for it. Some aspects of exploiting the potential of the limited water supply include reducing waste, distributing water better, giving priority to activities that respond best to irrigation at times of annual water shortage, improving drainage so that no salting or waterlogging occurs, and being able to recycle some of the drainage water. Pumps, piping, spray jets, drips, and land levelling and contouring improvements all justify the costs. Timely (early) application of water – that is, before plants are stressed – is invaluable. On many farms, more efficient use of water, reduction in labour, higher yields and greater areas cropped can result from applying new developments in engineering and irrigation technology for cropping. The payoff can be quite high.

**Cropping constraints**

Evaluate circumstances that might act against (restrain) the achievement of cropping potential, such as weed control and weed resistance to herbicides, soil structure, and timeliness of operations. Factors restraining (or preventing) the farm from achieving its full potential could be uncertainty about how well a new crop grows, the market for the crop, the labour needs, the new mixture of crops not satisfying the full needs of a rotation, no suitable land for development, or the cost as well as the difficulty of development being so great that it wouldn't be worthwhile to plant the crop.

**Fertiliser constraints.** Some farmers lack knowledge of the finer (but crucial) points on how to use and when to apply chemical sprays, especially for weeds, for best results. The main restraint on organic fertilisers to improve soil fertility is the small supply available to plants, compared with the large amount demanded by 'fertility-mining' cropping systems.

**Labour in cropping.** Sometimes, in cropping, there are peaks of demand for workers but there are not enough available hands to do the job on time. There are many possible arrangements for getting the job done. For example, farmers sometimes agree to share labour. Farmers can also hire workers. In between these extremes there are share-farming, renting and bonus deals, payment of which relies on crop yield and price. The potential for improving either the timing of the job or the standard (quality) of operation is worth investigation.

**Extending the length of the cropping phase in the rotation or sequence.** Since cropping is usually more profitable per hectare than the grazing component, it might be technically possible, through the use of additional fertiliser and herbicides, to extend the length of the cropping phase in a legume–pasture–crop rotation or sequence and still maintain fertility.

**Irrigation development.** Some restraints on developing irrigation are geological and hydrological. Other restraints include lack of capital either to develop more irrigation land or to buy the equipment to distribute the water properly, unreliable distribution systems from the main source of supply, and lack of knowledge about new developments in irrigation technology.

**Present machinery services**

When analysing how the farmer uses machinery, the question isn't 'What machinery equipment do you own?' but 'From what sources do you get your machinery services?' These services come from many places: from equipment the farmer owns (solely or jointly); from contractors; from share-farmers; through one of the many arrangements with other farmers; hiring or leasing; and possibly from other sources. These are the main points to be covered in analysing the present machinery services of a farm:

- how the machinery services are provided;
- whether the machinery services are available to do each operation properly and on time;
- the age, current value, and the anticipated timing and cost of a replacement;
- overhead and operating costs of machines;
- whether the implements match the tractors;
- work rates;
- how well machinery is being maintained;
- what equipment and facilities for maintaining it are needed (and available);
- availability of spare parts and repair services; and
- whether the machine is the right size and design for the range and the size of the jobs it has to do.

**Machinery potential**

When considering the future needs of machinery on the farm, it is necessary to know such things as:

- whether the overhead costs can be reduced by, say, providing some contract services;
- whether a change in the form of acquiring machinery services would be profitable;
- what measures could be taken to improve either the standard or the timeliness of operations;
- if it would be profitable to replace labour with more or larger machines;
- given that increases in machinery costs are likely to continue, where the capital will be found to replace machinery when the time comes; and
- if the power source and the implements can be better matched.

The areas that offer greatest potential for improving 'owned' machinery services are matching of capability and requirements, better maintenance to prolong the life of machines, and correct adjustment of equipment to enable the machine to do its job properly. The ingenuity of some agricultural engineers and the many farmer-inventors results in many simple modifications to existing machines, and better-designed machines.

**Machinery constraints**

Constraints on the efficient acquisition of machinery services are to do with size of operation, the availability of contract services, the quality and timeliness of contract services, and the availability of skilled operators and repair and maintenance services.

## Economic and financial conditions

The most interesting aspects of a business system are the economic benefits and costs associated with that system and the net benefits deriving from it, the financial aspects, and prospects for growth of the system and of the net worth of the owners. 'Economics' refers to the return on capital. This is the profit produced by the mix of the resources of land, labour, capital and management, and measures the efficiency of the use of all capital in the business. 'Finance' refers to whether there is the cash to pay the bills when they occur and service all the debts – that is, liquidity. 'Growth' means growth of wealth.

In assessing the economic and financial state (what is) and prospects (what could be) of a farm business, monetary values are placed on as many benefits and costs as can be done, in order to compare benefits and costs and estimate net benefits. The aim is that the most likely benefits exceed the most likely costs, and the cash flow is adequate to the task. Assessing the present situation and potential of a farm business involves identifying the inputs and outputs of a farm system and organising them into a benefit-cost (profit) framework and a net cash flow framework. The first of the important farm management questions identified at the start of this chapter was about how efficient (profitable) and liquid (financial) the business is. An example of how these questions are answered follows.

## Operating profit and return on total capital

Suppose a farm business has an expected gross income of $2m over a time period, usually a year. This comes mostly from sale of production, plus the value of some production that has not yet been sold but which is counted anyway because resources have been used (costs) to produce it in the time period in question. To produce this income there were direct costs, also called variable costs, such as fertiliser, fuel and chemicals used. The amount of these types of costs varies directly with the amount of output produced.

Gross income minus variable costs leaves total gross margin (TGM). If total variable costs were $600,000, then TGM is $1.4m. From this we must deduct all the other costs that don't vary regardless of how much production occurs. These are called fixed costs (or overhead costs), and include permanent labour, owner-operator's labour and management, rates on land, and depreciation of machinery and equipment from aging.

The TGM minus fixed costs leaves operating profit. If fixed costs are $400,000, then operating profit is $1m. This operating profit doesn't include finance or lease charges – these are costs of acquiring the services of the assets used but are not directly related to production.

First we want to know how much profit all the resources – land, labour and capital – have earned. If the total value of the assets (land, livestock and machinery) was $10m, then return on total capital is $1m/$10m, or 10% p.a. This figure is the measure of efficiency of capital used in the business, and the 10% p.a. return can be compared to what might be earned in other ways of using the same capital.

We also need to think about whether the value of the capital has increased over the year. If so, this is another form of return and has to be counted as well. In this case, the value of the capital invested in the business has not changed. (Even though machinery has lost value from depreciation, an allowance for this has been made.)

### *Net profit and return on owner's capital*

Commonly, the business will use capital invested by the owner and capital obtained from others. Then, those who have provided capital to the business have to be given a return, and they get the first share of the operating profit. The return to providers of capital is interest on borrowed capital, and lease payments on leased capital. Once interest and lease payments are deducted from operating profit, net profit before tax remains. Net profit is the return on the owner's capital. It indicates the efficiency of the owner's capital. However, efficiency of all the capital is the first concern in a farm business.

In this case, operating profit was $1m, or 10% p.a. of total capital, and there was no debt or leased capital, so net profit too is $1m, or 10% p.a. of owner's capital. Tax is yet to come out of this $1m net profit before tax.

### *Net cash flow*

On the financial side of the business, the quantity and timing of cash in and out is critical. The particularly important measure is cash in minus cash out, before interest and principal required to service debt. Then, net cash flow after debt servicing requirements is estimated. Suppose, in the case above, the gross income is all cash income and $2m cash in is expected through the year. The $600,000 variable costs are all cash, and cash fixed costs are $300,000 ($100,000 of fixed costs was machinery depreciation, not a cash cost). The cash fixed costs include permanent labour and manager's salary. As well, $100,000 is set aside to replace the machinery when it is needed, and a further $200,000 is invested in pasture improvement that will lift income in the coming years. Farm net cash flow before debt servicing is $2m – $1.2m = $800,000. If there were interest and principal repayments to make, there is expected to be $800,000 available.

The focus in farm business management ultimately is on the future of the business, even though, to inform the decision, the past is relevant. The past determines the present, and the present is the starting point for the future that the business will eventually experience.

## Growth

The focus on growth of farm systems encapsulates, first, personal growth of the people in it; the continual learning necessary to retain mastery of the farm business system in a future that will change from the present situation which the farmer has mastered, to varying degrees. The future is a different world; they will do things differently there . . .! Second, the focus on growth is on business growth, which comes down to increasing net worth of the owners. Farmers might think of growth in terms of controlling more assets (for instance, land), or increasing sales of output as a result of intensifying and/or extensifying, or horizontal integration, or vertical integration, or diversifying into different activities. These are all legitimate views about the meaning of growth, but growth in wealth or net worth is usually the ultimate aim.

Increase in net worth is an increase in the wealth of the owner of a business, measured as the value of capital that would remain in the hands of the owners if all assets were sold and all debts paid. Increasing net worth is almost invariably an important objective of the owners of farm businesses. On occasions, the situation can be such that the owners are at a stage of life where they are able and content to maintain or consume accumulated net worth. However, for most of the careers of most owners of farm businesses, growth is an imperative and no-growth isn't an option. This is because the imperatives of competition from fellow farmers dictate that if the farm business doesn't grow over time, it will inevitably not be viable sometime in the future. The farmer's saying is: 'If you're standing still, you're going backwards.' An example of assessing growth, and a demonstration of the critical role of using enough, but not too much, of other people's capital to enable the value of your own capital to grow sufficiently rapidly, is given below.

### Growth of owner's capital, equity or net worth

In the previous example, net profit was 10% p.a. return on total capital of $10m. If net profit after tax isn't consumed by the owner, such as spending it all on having a good time, then the net profit after tax is the amount by which the owner's own capital, called net worth or equity, has grown. This is what is meant by growth. This is usually an important objective of the owners.

If this business with $10m total capital had a $2m debt, the equity percentage is ($10m – $2m)/$10m, or 80%. If the interest rate was 8% p.a., an interest charge of 8% p.a. of $2m would have to be paid. This comes to $160,000. Net profit would then be $1m – $160,000 – $740,000. Return on own capital would be $740,000/$8m, or 9.25% p.a. With no tax or consumption out of this sum, the growth rate of equity would be 9.25% p.a.

### *Gearing and growth*

Gearing means the ratio of debt to equity that makes up total capital. The point to note from the above example is that the growth rate has increased by having some debt in the business. Other people's money cost 8% p.a. to borrow and was used to earn 10% p.a. in the business. This enabled the owner's capital to grow more rapidly than was the case when he relied solely on his own resources. This effect is called the benefit of gearing. Use other people's capital well and your own capital will grow more rapidly than if you relied only on your own capital.

However, this seems too good to be true. All a business person needs to do is borrow, use the borrowed capital profitably, and grow wealthy! This *is*, of course, too good to be true. If by chance the business was to make an operating loss amounting to 10% p.a. of total capital, instead of an operating profit, a loss of $1m would be made. As well, though, interest has to be paid on the $2m debt – 8% on a $2m debt comes to $160,000 in interest. Total net loss becomes $1.16m. This represents a loss of 14.5% of owner's capital. The rate at which owner's net worth has declined (14.5%) is greater than the rate at which it would have grown (9.25%) when the equivalent-sized profit was made. The downside effect exceeds the upside effect on net worth. This example demonstrates one of the main principles of business, the principle of increasing risk.

If this principle of increasing risk didn't operate, then businesses would be able to simply keep gearing up, borrowing more and more, and having their net worth continually growing. The world doesn't work like this, especially agriculture, because, first, income and costs vary greatly from year to year, and in any run of years there will be some profitable years and some unprofitable ones. As interest always has to be paid, in the unprofitable years equity is eroded more than it grows in the profitable years. Further, as the amount of debt in the business grows, the interest bill increases and the chance of making a loss increases, so lenders charge a higher interest rate, further increasing the size of the loss when it occurs. And, as a further blow, the reason for profit being down, such as a fall in prices, affects the value of the assets used in the business. So, where we had, say, 80% equity and 20% debt, if asset values halved from $10m to $5m, the debt remains at $2m and equity is now $3m, or 60%. The business is at more risk than before. Lenders' capital is also at more risk, and they accordingly charge higher interest rates, increasing the chance of a loss, and so on. The owner's equity is in a downward spiral.

The lesson is that there is a trade-off between the amount of equity and debt (called gearing), returns and growth in equity, and risk of losing equity.

## Business health: Return to capital, growth and debt servicing ability

The whole farm approach to analysing farm businesses refers to first considering all the elements – human, technical, economic, financial, risk, institutional – that affect the performance of a farm business. Second, it means looking at the business, not just in terms of the individual activities that make up the business, such as wheat, first cross lambs, or beef, but in terms of the performance of the whole business. This means estimating the total value of the assets and debts at the start of the year, the expected whole farm profit and net cash flow, and efficiency, using return on total and owner's capital, net cash flow before and after debt servicing, growth in owner's net worth, and value of assets and debts at the end of the year. This is what is called a business health check.

> Suppose at the start of the year a farm business had 2,000 hectares of land worth $2,000/ha (2,000 × $2,000 = $4m), livestock worth $1m, and machinery and equipment worth $500,000. Total assets are $5.5m. Debt is $1m – a 10-year loan just taken out, at an interest charge of 10% p.a. and with the $1m principal repayable in 10 equal annual instalments of $100,000 (called a term loan). The net worth of the owners is $5.5m assets minus $1m debt, giving $4.5m equity or net worth.
>
> Over the coming production year, the livestock – 20,000 dry sheep equivalent (DSE is the feed required over a year by a 45 kilogram wether; it is a standard livestock unit used to rate the annual feed demand of other animals, or feed supply of a pasture) – earn a gross margin per DSE of $20, giving a total gross margin of $400,000. Overhead costs are $200,000. Operating profit before tax is estimated to be: TGM $400,000 minus $200,000 overheads, which leaves $200,000 operating profit before tax. Return on total capital is $200,000 from $5.5m capital; a return of 3.6% p.a. on the capital invested.
>
> The return on the owners' capital ($4.5m) is $200,000 operating profit minus $100,000 return on borrowed capital; leaving $100,000

**Table 2.1** Profit, cash and growth

| Profit | | Cash flow | |
|---|---|---|---|
| TGM | 400,000 | Cash variable costs | 400,000 |
| Overheads | 200,000 | Cash overheads | 125,000 |
| Operating profit | 200,000 | Machinery replacement allowance | 75,000 |
| Interest | 100,000 | Interest | 100,000 |
| Net profit | 100,000 | Principal | 100,000 |
| Income tax | 15,000 | Income tax | 15,000 |
| Growth in equity | 85,000 | Net cash flow | –15,000 |
| Balance sheet at end of year | | | |
| Assets | | Debts | |
| Land | 4,000,000 | Original debt remaining | 900,000 |
| Livestock | 1,000,000 | New debt (cash deficit for year) | 15,000 |
| Machinery plus plant | 425,000 | | |
| Machinery replacement allowance | 75,000 | | |
| Total | 5,500,000 | | 915,000 |

Equity: $4,585,000.
Growth in equity: equity at end is $4,585,000 minus equity at start $4,500,000, a growth of $85,000.

return on owners' capital (called net profit, or net-farm income). From the $100,000 net profit comes $15,000 in income tax. The remaining $85,000 belongs to the owners, to do with it what they wish. If they save it, or use it to pay off a debt, or to buy a new asset, the owners' equity increases by $85,000. The $85,000 represents 1.88% growth in equity ($85,000 as a fraction of $4.5m equity at the start of the year).

Looking at what has happened during the year from the viewpoint of cash flows, another story emerges. If livestock activity gross margin was all cash, then cash income minus cash variable costs leaves $400,000 cash gross margin. Not all of the $200,000 overhead cost is cash – depreciation of plant and machinery at, say, 15% of the value of $500,000 is $75,000 depreciation cost that isn't a cash cost. Suppose, though, that $75,000 is set aside in a bank deposit as a machinery and plant replacement allowance. The cash flow budget will be $400,000 TGM minus $125,000 cash overheads minus $75,000 cash set aside for machinery replacement minus $100,000 interest payment minus $100,000 principal repayment of the debt minus $15,000 income tax. This leaves a net cash flow of –$15,000. The business has made $200,000 operating profit before tax and a net profit before tax of $100,000, and has a net cash flow of –$15,000 after meeting all obligations. The balance sheet at the end is different from the balance sheet at the start. Why? Assets have increased – there is a $75,000 bank account for machinery replacement, and $100,000 of debt has been repaid. But machinery and plant assets are worth $75,000 less because they have depreciated in value, and a new debt of $15,000 has been incurred. The overall change in the balance sheet is an increase in net worth of $85,000.

The situation can be summarised as set out in Table 2.1 (p. 35).

## Investment

Analysis of investment is about analysing the expected stream of benefits and costs from the assets involved. When investors buy assets, they expect to incur initial capital costs and subsequent operating costs that make it possible to receive future benefits in the form of annual after-tax profit and capital gain. They want the sum of operating profit and capital gain to represent a percentage return after tax on the capital invested that is comparable to alternative opportunities to invest, with an allowance for the relative risks involved and for non-pecuniary benefits they may derive from being in this type of business and not another.

A key idea in the theory of investment is that if there were no differences between assets in terms of profitability, liquidity, taxation effects and risk, and if investors were equally well informed and competitive and had similar expectations, then the expected returns on all assets and the value of the assets would be the same. This is because investors would bid the price of an asset up to a price at which it yields the same returns as elsewhere. In reality, expected returns per dollar invested for different assets differ markedly. This is because the important attributes of an investment – expected returns, riskiness, treatment under tax law and liquidity – are different for different assets. So, when judging an investment, the focus is on expected returns, riskiness, tax and liquidity. The relationship between risk and return is that investors can only get higher returns by taking more risk. There is no other way: low risk, low return; high risk, high return; average risk, average return. To reduce risk to below average is to eliminate the chance of above-average profits.

Another key idea in investment theory is the efficient market idea, which holds (to varying degrees) that at any time, all available information is fully reflected in the price of an asset. In such a situation, changes in the price of an asset are the result of unanticipated changes in factors that affect the value of the assets. Thus, changes in asset prices must be random. When economists refer to efficient markets, they are referring to one in which relevant information is widely known and quickly distributed to all participants. Theoretically, average return, risk, tax treatments, and so on, of investment alternatives will be fully known by all investors. Then, as the participants in this market have all the information, the market prices of assets reflect this information. For markets to be efficient doesn't require that everyone involved in the markets has all the relevant information. All that is needed is that *enough* participants have the information – in this situation, prices will move as though the whole market had the information. Prices will be quickly bid up or down to levels that reflect the complete information. Uninformed buyers, buying at current prices, will get the benefit. As the saying goes: 'Investors cannot beat an efficient market; they can only be lucky in it.' For example, in trying to make profits by trading in an efficient share market it isn't enough simply to choose companies that are expected to be successful in the future. This is because everyone else with the same information will expect them to do so too, based on the available information. The prices of shares in those companies will already be quite high. In the situation where investors are well informed, the way to make profits from share trading is to pick companies that will surprise the market by doing better than is generally expected. The investor needs to invest in those companies before the expectations of other investors in the market changes – and so does the price.

The market reflects an average value of people's estimates about how much a share in a company is going to be worth, and this reflects how well they expect the share to perform in the future. Tomorrow's news is, by definition, unanticipated. No one can predict accurately whether it will cause the share price to rise or fall. Therefore, in an efficient stock market, prices will move unpredictably, depending on unexpected news. The way to do better than the market returns on average is to take greater risks than that of the average investor. And taking greater risks means there is a larger chance of doing much worse than the average of the market, too. The exception to efficient market theory is insider trading. This is where people on the inside have information that they exploit to their advantage.

In sum: As prices in an efficient market already reflect all the available information, price changes in such markets are responses to unexpected news, because if the event had already been anticipated its expected effects would already be included in the price: 'If the future was known with certainty, it would have already happened.' The theory of efficient markets says that most investors cannot beat the market – they cannot earn better returns from their investments than the average of the market unless they have information that other investors don't have. The psychology of the market means that people are not betting on what they think the market will do. They are betting on what other people are likely to think the market will do. This psychology can shift in unpredictable ways: one word for this is confidence. Volatility in confidence and expectations is the explanation for changes that occur in the market price of shares that are not related to any identifiable real-world phenomenon.

An example follows that shows how farmers can answer the question about investing more capital in their businesses.

## Investing capital to improve whole farm profit

A farmer is considering investing $15,000 on 200 hectares of annual fodder crop in a paddock that would otherwise have produced only enough feed over the year for 1,000 DSEs with a gross margin of $15/DSE – that is, a total gross margin of $15,000. By sowing the paddock to the fodder crop, the farm will be able to carry an extra 2,000 DSEs, giving a total of 3,000 DSEs, at $15/DSE, a TGM of $45,000. This is a gain of $30,000 over leaving things as they were. The extra animal production costs are already accounted for in the gross margin/DSE of $15, and the other extra annual cost (before extra tax and extra interest) is $15,000 to grow the fodder crop.

The gain before tax and extra interest is extra gross margin of $30,000 minus extra annual cost of $15,000 = $15,000 gain. The extra capital invested is 2,000 DSE × $20/DSE = $40,000. The return on extra capital before extra tax and interest is $15,000/$40,000 = 37.5%.

If extra tax payable is $3,000, return after tax is $12,000, a return on capital of $12,000/$40,000 = 30% p.a.

This can be compared to the cost of capital. What is the cost of extra capital invested in this case? The cost of extra capital invested is a mix of the cost of the extra capital invested that is borrowed, and the opportunity cost of the owner's own capital that is put into the project. The cost of the borrowed capital is the market rate of interest charged on the borrowings. The cost of the owner's own capital invested in the project is the returns forgone by not investing it elsewhere (the opportunity cost). If the owner could invest safely at 6% return on capital p.a., or in a similarly risky environment and earn 12% on capital p.a., then the owner would consider the cost of investing in this project, which has quite a bit of risk, to be 12%. If half the $40,000 total capital is borrowed at 8% p.a. and the other half is the owner's own capital at a cost of 12% p.a., then the cost of capital (called weighted average cost of capital) is $0.5 \times 8\% + 0.5 \times 12\% = 10\%$ cost of capital p.a.

## Investing in extra land for a wool-producing property

This farm business comprises 640 hectares carrying 6,000 DSEs as sheep for wool, in average to good times earning a total gross margin of $120,000 with cash overheads of $35,000, allowance of $45,000 for operator's labour and management, $10,000 for depreciation and an operating profit of $30,000. Total capital invested is $1.2m. The farmer thought that annual nominal return on capital after tax had probably fluctuated over the years by around 3%, sometimes zero, sometimes 4–5%.

The owners of the farm were aged in their early forties and wanted to continue farming for another 20 years. To do so, they realised they would have to expand their operation whenever a reasonable opportunity arose. Their children were not going into farming. A 200-hectare block of land approximately 3 kilometres away was for sale by auction.

These farmers had 97% equity in their business, and had accumulated assets off the farm of $400,000 that they were prepared to use in purchasing more farmland. The land for sale was of poorer soils, aspect and topography than the current farm, had a poor fertiliser history, and had been poorly managed for the past 15–20 years.

These farmers carried out a standard preliminary farm management analysis – that is, they estimated the likely extra returns, extra costs and extra return on extra capital per annum for the situation when the change was implemented and when it was operating fully (called a steady state). From the results of their analysis they believed that the current carrying capacity of the 200 hectares of farmland was 6.25 DSE per hectare, 1,250 DSE in total. The stock capital required would be $20,000.

These farmers reasoned that without further investment, the property could generate a total gross margin of $25,000, or $18,000 net after extra tax and rates, extra transport and repair and maintenance expenses, depreciation costs and, possibly, a small amount of extra casual labour.

With development, using well-established methods that worked in the district, the farmers judged that the property could nearly double its current carrying capacity to around 2,000 DSE. This would involve a capital investment of $150 per hectare, a total of $30,000, plus $20,000 in livestock. The extra return would be from 800 extra DSE, at a gross margin of $20 per DSE, a total of $16,000 extra return. This would reduce to $13,000 typically, after some allowance for extra costs of risk associated with the increase in stocking rate, such as buying extra feed in the inevitable dry years. With these figures, the real after-tax return on marginal capital from development of the land once purchased would be around 20% p.a. This was attractive to the owners, exceeding any other investments they could identify. Thus, they decided to try and buy the land with the aim of developing it to increase its carrying capacity.

The first-look analysis showed that at present carrying capacity, and an expected steady state net return per hectare of around $90, a 6% required rate of return would make the land and stock worth around $1,500 per hectare – that is, $90/0.06 = $1,500, or $90/$1,500 = 6%. Given the potential for development of the property, the farmers' judgment was that if they could get the land for $300,000 ($1,500/ha × 200 ha) including transaction costs, plus invest another $30,000 for development to carry 800 extra DSE and $20,000 for extra stock, it would be a good land purchase decision. Further, at this price they could finance the investment using their own resources. They went to auction and purchased the land for a price close to $1,500 per hectare.

The farming couple based their purchase price of the land on two main pieces of information:

- expected net returns at current carrying capacity; and
- expected extra return to their extra capital after development to increase carrying capacity.

They also ensured that they were able to finance the investment using their own resources. They were not prepared to take on too much risk and, while keen to expand their land holdings, were also cautious in the way they did so.

## Investing to acquire machinery services

A crop farmer might weigh up the choice between purchasing alternative headers, or acquiring the machinery services in another way. In this situation the farmer needs to account for the following costs:

- cost of owning the machinery;
- cost of operating the machinery;
- costs that might be incurred because of lack of timeliness or quality of operation with some machines compared with others of various sizes; and
- cost if alternative sources of acquiring machinery services are employed, such as share ownership or contracting services. And, if an alternative to ownership is used, are there any related implications for the whole farm system, such as labour savings or timeliness, quality and autonomy effects?

In another case the farmer might face only the choice of owning one type of machine or another machine. The ownership and operating costs can be estimated on an annual or lifetime basis. On an annual basis, the costs are expected to be:

| | |
|---|---|
| Header purchase cost | \$300,000 |
| Expected life | 10 years |
| Expected salvage value in year 10 (in today's \$) | \$50,000 |
| Expected annual depreciation (\$300,000 – \$50,000) ÷ 10 | \$25,000 |

### *Annual opportunity interest cost on capital invested*

The value of capital tied up in the machine will differ each year of the machine's life. To handle this, we estimate the average amount of capital invested over the life of the machine. In this case, the average capital invested is estimated as:

$$\frac{\text{Opening value} + \text{Closing value}}{2} = \frac{300{,}000 + 50{,}000}{2} = \$175{,}000$$

The opportunity interest cost charged is a real interest rate (the whole analysis is in real dollars and real interest rates, where 'real' means with no effects of inflation included), say 5% real = \$175,000 × 0.05 = \$8,750.

The annual depreciation and interest cost is $25,000 + $8,750 = $33,750.

As well, there will be annual registration and insurance, and sometimes a shedding cost.

Annual operating costs are straightforward: batteries, tyres, fuel, oil, repair and maintenance, and casual labour if needed.

Suppose the total annual cost of owning and operating a header was:

| | |
|---|---|
| Depreciation and interest | $33,750 |
| Registration and insurance | $1,000 |
| Operating cost/hour with hired casual labour to drive header | $56 |
| Hectares worked/hour | 8 |
| Contract rates/hectare | $20 |

To decide whether it would be most economic to own the header or to use a contract harvester, the following calculation is done:

Ownership cost: $33,750 + $1,000 + Operating cost/ha ($56 per hour/8 ha per hour, $7/ha) × no. of hectares = Contract cost of $20/ha. This becomes (call the no. of hectares '*x*'):

$$\$33{,}750 + \$1{,}000 + \$7x = 20x$$
$$\$34{,}750 = \$13x$$
$$2{,}673 \text{ ha} = x$$

If the farmer has 2,673 hectares or more to harvest per year, it is cheaper on a cost per hectare harvested basis to own the header. If the area to be harvested each year is less than this, it is best to use a contractor. As a check on this, if 2,672 hectares were harvested using a contractor, the total cost is 2,672 ha × $20/ha = $53,400. Owning the header, the same area harvested would cost 2,672 ha × $7/ha + $34,750 = $53,454. Owning the header and harvesting, say, 3,000 hectares would cost $55,750. A contractor would charge $60,000 to harvest the same area. Note: This is without explicit values being placed on the important criteria of timeliness and quality of operations. If owning the machine also meant benefits of, say, $20,000 from improved timeliness and quality of output, then the breakeven area that justifies owning the machine is reduced to 1,673 hectares ($53,454 cost of owning machine minus

$20,000 gain in timeliness and quality, or cost of untimeliness and lower quality avoided/$20/ha contract charge).

The main criteria used to judge investments are: (a) *economic:* expected extra return on the extra capital invested, compared with what the same capital could earn in an alternative investment; and (b) *financial:* whether the funds required for the investment can be borrowed and interest paid and borrowings repaid. Two sound rules for thinking about investments are:

- evaluate the return, risk, liquidity and tax aspects of the investment; and
- aim for a portfolio of investments that has a range of risk attributes.

The risks associated with any proposed investment have to be included in the evaluation of expected return on capital. Many apparently good ideas are not implemented because, once risk is taken properly into account, the project isn't worth doing. The expected returns are not high enough to cover the risk that it might fail, because of technical failures or unfavourable climatic conditions or market prices, and cause unacceptable outcomes.

## Risk and uncertainty

The 21st century is anticipated to be the age of understanding risk and uncertainty – at least, much more so than at present (Bernstein 1996; Shiller 2003). Economics as the discipline about making choices inevitably also is the discipline of risk and uncertainty. Risk and uncertainty have long been a major worry to investors and analysts alike – and yet, intriguingly, it is precisely the existence of risk and uncertainty that creates the opportunities and rewards that people are in business to capture. So, in the analysis of decisions, both offensive (opportunities, profits) views of risk, as well as defensive views of risk and uncertainty as something whose consequences are always something to be reduced or avoided, have a role.

In the economy there are many ways in which the risks faced by farm businesses are changed or shared or transferred or removed. Risk is in part a consequence of decisions made within the business and is generally divided into two types: business risk and financial risk. Business risk is the variability in the returns on assets that is affected by the business's investment and operational decisions. Financial risk is more specific. It is the increase in variability in returns to the business owners as a result of using debt. The amount of debt used by a business affects the cost of debt – financing requirements become larger, and the cost of extra finances increases.

The differences between a bad decision and a good decision, and a right decision and a wrong decision, are fundamental to understanding the role of risk in farm management decision making. Bad decisions are those that have no hope of successful outcomes even if favourable conditions prevailed. A good decision is one that is based on the best information and judgment available at the time the decision is made. Whether it turns out to be the right decision or not will depend on the outcome of the events – that is, it depends on whether the horse wins or not! Good fortune has a role to play in success or failure in farming – or in any other activity, for that matter. Good luck (remembering that the better prepared people are, the 'luckier' they usually are), good timing (and price) of land purchase and development decisions, and good management – all these factors play big roles in success and survival in farming. The best way to continue to be a farmer is to avoid making bad decisions about buying or developing land, and acquiring the services of machinery; good decisions that turn out to be wrong decisions are hard enough to live with!

Managing a risky business is about gathering relevant information, weighing it judiciously and acting accordingly. When a change is made on a farm, the final outcome won't be precisely that which is thought most likely to happen at the time the change is made. Numerous different outcomes are possible. Analysing changes and making decisions requires taking account of the volatility of potential outcomes. There are ways of doing this formally. First, there is the use of the strengths of beliefs of decision makers about the likelihood of things turning out one way or another – called estimates of probabilities. Second, there is the use of information from investigation of 'what if?' scenarios using budgets (sketches or models) built into the computer-based spreadsheet, and weighed up in terms of how likely they are to happen. Less formally, decision makers use their intuition (also known as 'gut-feeling' and judgment). Each of these approaches is useful when weighing up the merit of making a change on a farm. Using approaches to decision making that are more structured than acting solely on intuition and lessons from past experience, or than acting randomly, can give a better chance of making the right decision. This involves making sure that all relevant data is considered, and that a consistent and logical process is used to select the action, or set of actions, most likely to reach the goal(s). These goals have to be defined clearly.

It is overly simplistic to reduce decision analysis to analysis of 'once and for all options'. Sound farm management analysis isn't about conducting more and more analyses of a single decision; it is about having a reasonable go at analysing the usually small number of potentially plausible choices, taking a decision and making it work, and responding as the farming world changes both as a consequence of the decision and independently of the decision. In practice, sequences of decisions create the future, and eventually the history, of any business.

Decision makers, faced with risk and uncertainty, are constrained by what is known or knowable or imaginable. It is analytically useful to distinguish between types of uncertainty. First, there is uncertainty caused by the unpredictability of the future and of future events. Second, there is uncertainty in the analysis caused by the unknowability of certain things. Many variables cannot be estimated at reasonable cost. The question becomes: how can uncertainty be handled in analysis? First, uncertainty has to be recognised – this helps to give a feel for the quality of the information included in the analysis and therefore the faith that can be placed in the conclusions.

The ultimate role of risk in decision analysis is in part determined by all of the risk-related tactics and strategies used and the whole business situation established by farmers. The notion of risk as a commodity to be sold by those wishing to reduce their exposure to risk, to be bought by those willing to accept more risk and consequently earn higher average returns, is truly profound. The main focus of decision analysis then ought to be about opportunities offered by markets to trade the risk and uncertainty associated with changes to farm systems. (The constraints imposed by a relative scarcity of liquidity in the face of growing markets for risk is another matter.)

'Life is risky. We can't remember the future' (Anderson and Dillon 1992). As we cannot remember the future, the key to understanding and managing risk is to imagine the future instead: imagining a small number of state(s) of the farm business if a small number of events and situations occur simultaneously, and relatedly. Exploring the consequences for goals of a small number of discrete scenarios encapsulating significant combinations of events, both sequential and simultaneous, provides useful information to decision makers.

The biggest challenge a farm business person has to deal with is that the future cannot be known with certainty. Things might happen – but then again, they might not. The good thing about this, though, is that this risk and uncertainty makes it possible to earn good profits. If there isn't much risk, there isn't much profit either. The tricky balancing act is to set the business up with enough exposure to risk to achieve the profit and growth goals without at the same time jeopardising the achievement of goals to do with staying in business and not going bankrupt, and enjoying being in the business. (As the old farmer said to his favourite horse: 'It's not old age that is killing you and me, Nugget, but hard work and worry, worry, worry!') There are some common-sense rules about managing risk to achieve important goals. These include:

- Understand well the risk–return tradeoff.
- Take a whole of business view of the risk situation, and manipulate the whole range of risk factors to establish an overall risk position that you can live with.
- Be geared up – but not too much.
- Capital on farm has to be used highly efficiently. The starting point is to be technically very good at farming.
- Get some capital invested elsewhere in an activity whose volatility has no correlation with the volatility of agriculture.
- Sell a proportion of the most critical risks to someone else, using forward and futures contracts to remove some of the volatility from some of interest rates and prices and costs.
- Remember that by reducing exposure to risk, you reduce profit as well as loss.

An example of how a manager of a farm business might consider risk follows.

> Suppose a farmer is weighing up a potential investment in further development of the existing farm that would involve $200,000 capital investment, with a life of 10 years, and with all the initial capital redeemed at the end of the life of the project. Alternatively, a similar amount of capital could be invested elsewhere, off the farm, where it would be highly likely to earn 8% on capital after tax on average over each of the next 10 years, with all the initial capital redeemed at the end of the life of the investment. The farmer feels that there is a very good chance that the 8% will actually be earned on average over the life of the non-farm investment. The farmer needs to consider the returns and risks from the on-farm investment relative to the alternative. Suppose the farmer judges the riskiness of the farm investment as follows:
>
> - A best return of 30% after tax will occur one year in 10.
> - A good return of 15% after tax will occur two years in 10.
> - A satisfactory return of 12% will occur three years in 10.
> - A poor return of 6% after tax will occur two years in 10.
> - A worst return of –10% after tax will occur two years in 10.

The 'number of years in 10' that something is expected to occur can be converted into 'the odds' – as in betting on horse-racing, or probabilities. Something that is expected to occur two years in 10 has odds of occurring of 4/1 against – that is, eight chances of not occurring and two chances of occurring. In terms of probabilities, two chances in 10 is a probability of 2/10 = 0.2. (To convert odds to probability, the rule is 1/(1 + the odds), so 4/1 against is a probability of 1/(1 + 4) = 1/5 = 0.2.)

Thus in this 'investment race', the odds about the outcome in any one year are:

- 9/1 a 30% return (0.1 probability)
- 4/1 a 15% return (0.2)
- 7/3 a 12% return (0.3)
- 4/1 a 6% return (0.2)
- 4/1 a –10% return (–0.2)

But this investment race is going to be run 10 times over 10 years. What is the expected average outcome – called 'expected value' – over the whole period? The expected value is the sum of each of the identified possible outcomes multiplied by the probability of it happening. Thus:

0.1 × 30% = 3%
0.2 × 15% = 3%
0.3 × 12% = 3.6%
0.2 × 6% = 1.2%
0.2 × –10% = –2%

Sum = 8.8% (This is called the expected value of returns.)

The farmer can compare the expected value of the returns of 8.8% p.a. from the farm investment with the alternative 10-year investment of $200,000 off-farm that is expected to earn 8% p.a. This information is used along with other considerations, such as the balance of the farmer's overall investment portfolio and exposure to different parts of the economy, and a decision made accordingly. A structured way of thinking about the likelihood of how an investment might turn out can be more informative than intuition alone.

### *Tracking performance*

Most farms face uncertainty of such magnitude that year-to-year financial performance fluctuates a great deal. This means that it is unreliable, and downright dangerous, to use surrogates or indicators of performance. In other industries, managers can set, and assess performance against, variables that are very predictive of overall performance; variables such as total physical output, sales volumes, physical productivity and market share. These measures of performance correlate highly with overall financial outcomes, year in, year out.

In farming the extent of variability across physical outcomes, input costs and realised prices for output causes relationships to move around. Sometimes high yield is associated with high profitability, sometimes not; it depends on output prices – if all producers have high yield, prices may be poor. Sometimes high sales revenue is associated with high profitability, sometimes not; it depends on costs – high revenue may reflect low industry output due to drought, which will have driven up production costs.

The recipe of inputs, throughput and output that determines financial outcomes is unstable year-to-year. This means that it is important to monitor the core outcome measures of performance. Tracking indicators, rather than overall outcomes, is tempting but can be too much like extrapolating success in a game from the absolute quality of a single move: tactical brilliance in the midst of overall loss.

We emphasise the tracking of profit, return on equity, cash flow and profitability for these reasons, and because the daily focus on production so easily beguiles any manager into overstating, implicitly to themselves, the significance to overall performance of specific measurable output achievements. The greater the uncertainty any manager faces, the more important is it to watch progress of the game rather than the 'success' of specific moves.

## Returns on investment in farm and non-farm businesses

In agriculture the operating returns on capital vary considerably for any farm business over time and, even more so, between farm businesses. Similar applies to returns from capital gains. Farmland in areas influenced by non-commercial agricultural demand for farmland may experience sustained growth in real capital value. Commercial farms away from such influences are more likely to see real capital values maintained in real terms by farmers who maintain profits by continually increasing productivity, but with capital values growing very little. Increases in real capital value of commercial farmland result from increased real profitability of the land use. Often, in agriculture, farmers are doing a good job simply to increase productivity sufficiently to maintain real profits in the face of rising real costs and declining real prices for output.

In recent decades, in terms of operating profit before tax as a return on total capital (not capital gains) and in a period of low inflation, the top quarter of farm businesses have earned an average of around 7% on capital in nominal terms per annum over a run of years. Average farms have earned an average of 3–4% nominal on capital. A small proportion have battled to earn an average of 1–2% nominal operating profit as a percentage of total capital. At the same time, the land value tended to keep up with inflation, and in some cases to exceed it. Adding 3% inflation of asset values gives nominal returns on capital of 10% p.a. for the top farm businesses, 6–7% for average operations, and 4–5% for the lower-performing businesses. (Note that 10% nominal return on capital with 3% inflation means real gains of 10 – 3 = 7%.)

Other asset classes, such as government securities, listed property and the All Ordinaries could expect to earn an average of around 10% nominal p.a. from capital growth and dividends. In a study of the period from 1988 to 2003, Carroll (2004) found compound growth rates for non-farm investments to be as follows:

- All Industrial shares: 10.5%
- All Ordinaries shares: 9.6%
- listed property: 11.43%
- government bonds: 9.9%
- listed food and household businesses: 7.3%
- cash: 7%

For the same period, Carroll found that farm businesses in the top 25% of the performances in the whole distribution of all farms achieved a compound annual growth rate of 11.9%. The average farm business earned a compound annual growth rate of 6.3%.

The other major part of the returns on capital story is the risk associated with the venture. In the analysis by Carroll, the dispersion of returns around the median return over the 15 years (measured as the standard deviation of the returns) was estimated. The farm businesses in the top quarter of all farm performances had a compound annual growth return slightly more than did the shares making up the All Ordinaries share market index (11.9% versus 9.6%), and had a similar degree of volatility – that is, a dispersion around the mean of 16%. All

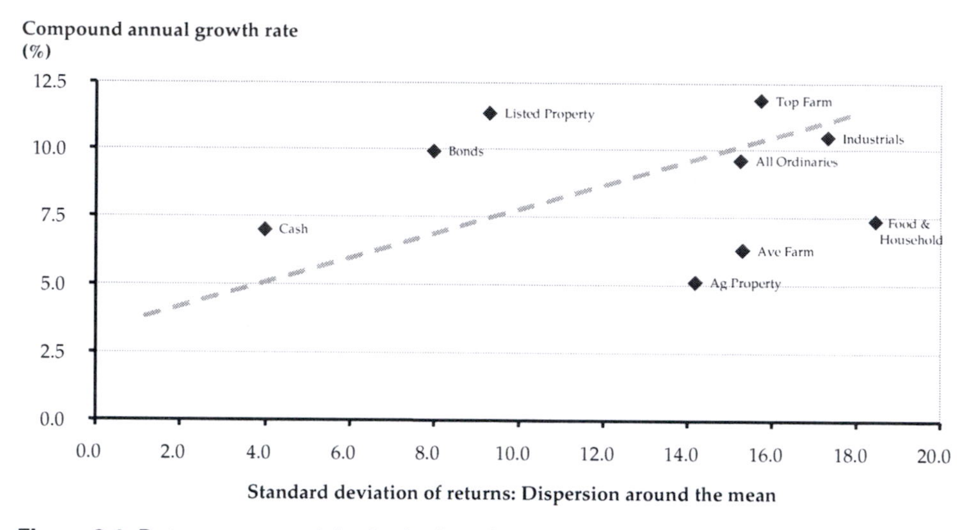


**Figure 2.1** Return versus risk: Agricultural and non-agricultural investment, 1988–2003
Source: Carroll (2004), *Farm Performance from a Wealth Creation Perspective*, Agribusiness, National Australia Bank.

Industrial shares had a compound annual growth return of 10.5% and a volatility of 17% variation around the mean over the 15 years. The average farm earned a compound annual growth rate of 6.3% with a volatility of 15%. Bonds and listed property earned at compound growth rates of around 10% and 11%, respectively, with volatility of 8% and 10%, respectively. Cash earned 7.5% with 4% volatility (see Figure 2.1).

Care is needed when interpreting estimates of performance such as those listed above, particularly in considering whether they have any meaning for the future. Mostly, such information is valuable as historical information that has little to say about the future, because the states of affairs in any period of time are unique to that time. The period 1988–2003 encompassed the usual range of farm and share market booms and busts, major drought, and a sustained period of low inflation and relatively high real interest rates, plus an $A exchange rate that ranged from $0.80 US to $0.47 US. The main point is the relationship between risk and returns – higher returns are associated with higher risk, but different investments with similar returns exhibited different risk, too. The message from this is that it can be possible in the economy to maintain similar overall returns on capital and reduce overall risk by establishing a diversified portfolio of investments with different risk profiles. The aim is to establish a portfolio of investments that encompasses a mix of returns and risk profiles, and has an overall return–risk mix that suits the investor according to their needs and objectives and personal make-up. If different investments have returns that are not affected by the same phenomenon to the same extent at the same time, and are not all good or all bad at the same time (called 100% positively correlated), or are such that when one is bad the other is good (called negatively correlated), then it becomes possible to combine investments in a mix that is likely to reduce exposure to risk while earning the returns required over time.

## THE ECONOMIC SYSTEM BEYOND THE FARM GATE

In this section, key principles are outlined about the behaviour of people and firms from beyond the farm that affect the way farm business systems operate. These principles are about people and firm behaviour and are, in many ways, fundamental 'rules' of the farming game.

### Firms and the economy

To understand how an economy works, it is useful to define what comprises an economy. Broadly, an economy consists of households, firms and a government sector. Households sell labour to firms and governments, and provide other resources, too, and earn income from so doing. They use their income to buy goods and services from firms and governments. Firms are organisations that combine resources of land, labour and capital to produce goods and services. All economies have governments, which are large producers of goods and services in their own right. They regulate, broadly and in detail where appropriate, what firms and households are able to do in the economy.

Resources in an economy can be categorised broadly as land, labour and capital, and management skill. These resources are limited relative to all the things that people could do with them if they were not limited. This is called the basic economic problem. There exist unlimited needs and wants of people and limited resources available to meet these needs and wants. If people are to be as well off as they could be from their ability to obtain control of resources and use them to meet as many of their needs and wants as possible, then they need to make the most from these resources. Getting the most out of the limited resources available and satisfying as many of people's needs and wants as can be done is called economic efficiency.

How are decisions made in an economy about the way resources are used? In the Australian economy, most of the decisions about the way resources are used are made by the private entrepreneurs who own and control most of the resources. In a market economy, entrepreneurs make decisions about the use of resources and production of goods and services in response to the messages they receive in markets about the costs of inputs and the value of outputs. Messages from the market come from the buyers and sellers of the goods and services that businesses use or provide. A market is a place where people make contact to buy and sell goods and services. In a market, buyers are continually making up their minds about how much of something to buy at the price at which it is offered and given its quality; at the same time, sellers are making up their minds about how much to sell at what price in light of what it costs them to make the product.

Market supply means the amount of a good or service that producers will provide in a market at various prices. Markets are said to be in balance, or equilibrium, when supply equals demand and the market clears. Market equilibrium would occur if every seller of goods or services (called supply) could find a buyer and every buyer (called demand) could find a seller. Supply from producers would equal demand from consumers. While many factors influence the amount supplied and demanded, the price of the good or service is the main factor and this is the focus of economists.

Firms in an economy operate in one of four major forms of industry structure. First there is the competitive industry structure. In competitive industries there are many producers and consumers; the competing firms produce products that are of similar nature and quality; no firm is large enough to influence market prices or affect what other firms need to do; and getting into or out of the industry is easy and not too costly. The opposite structure to competition is monopoly or monopsony. Under monopoly, there is only one seller of the good or service. Monopsony is where there is only one buyer. In these situations the single firm in the industry has power and can influence market prices.

The most common industry structure is one in between the two extremes of competition and monopoly, which is called oligopoly or monopolistic competition. Oligopoly is where firms have to consider what other firms in the same industry do because there are a reasonable number of firms producing similar goods and services, and actions by a firm in the industry will affect the other firms. Monopolistic competition is where firms try to gain some influence in the market by making their product somehow different from that of their closest competitors. These firms thus try to reduce the full impact of competing firms on their own sales. Monopolistic competition firms are said to differentiate their product in order to establish their own set of customers.

The central tenet of the competitive market model is that competition between firms to supply goods and services will force them to produce at the lowest cost per unit of output that is possible with their given size of operation. Theoretically, over time prices of goods and services will be brought down to this lowest possible cost because, when prices exceed this competitive minimum cost level, then firms will be making better profits than happens in other activities in the economy, and so new firms will enter the industry and increase supply and force prices down. Firms that cannot produce at the new lower prices will leave the industry. Thus, the theory goes, competition between firms will bring the product price down to the lowest possible level, and aims of using resources efficiently will be achieved.

The quantity and quality of goods and services that firms are prepared to supply represents the amount of a good or service that the firm can profitably put on the market at prevailing prices. The typical supply relationship is that as price increases, so too does output – with other things such as technology, prices of inputs to production, and prices of other products that could be made being held constant. There is a range of quantities of product that firms would be prepared to supply at a range of possible market prices. This is because, as output increases, usually the costs per unit of output increases and so higher output will only be supplied at higher market prices.

The demand of consumers for goods and services is represented as the relation between prices of the good or service and quantities, with all other influencing factors held constant. The other factors held constant include the income of the 'demander', the tastes, the number of demanders, and the price of related products. Thus, with these assumptions, the amount of a good or service consumers wish to buy is directly a result of the price of the good or service. An individual consumer's demand for a good or service tells how much the consumer would want to buy at a range of market prices, for a given set of circumstances such as level of income, prices of other related products, and so on. This relationship, known as the law of demand, shows that as prices of products increase consumers will buy less of a good or service, and as prices decrease consumers will buy more goods and services.

The theory of market supply and demand holds that consumers will demand more of a product or service at low prices than at high prices, and producers will supply more of a product or service at high prices than at low prices. Demand and supply represent the state of affairs that would exist in a market at a point in time with everything that can affect them, except price, held constant. It is like a snapshot in time, which shows that, as price changes, more or less of a good or service will be supplied or demanded. Once time is introduced into the question, then these assumptions are relaxed. Technology can change, tastes can change, income can change, population size can change, prices of other inputs and outputs can change, and prices of related products can change. The net effect is that supply and demand can shift once time is introduced and changes in other influences on producer and consumer behaviour are introduced. A shift in demand or supply means that, at a given price, there will now be a different amount demanded or supplied. Or, the same quantity as previously can now be supplied or demanded at a different price. For instance, a fall in the costs of producing a good or service, such as might happen with an improvement in technology, has the effect of enabling more of the good or service to be supplied at the existing price, or the same amount as was previously supplied can be supplied at a lower market price. In agriculture, farmers continually improve their efficiency and increase the supply of farm products at lower prices.

On the demand side a change in income, population, tastes, or prices of complementary or substitute goods or services all have the effect of changing the quantity demanded at the existing price. For example, a change in the price of an alternative product will have the effect that consumers will demand less of a product at the current market price, or will only buy the same amount as currently at a lower price.

An extremely useful idea is the amount by which quantity supplied or demanded changes as the price of the good or service changes. This is termed price responsiveness or price elasticity and is measured as the percentage change in quantity demanded or supplied for a given percentage change in the price. An interesting question to producers and consumers alike is: 'If the price goes up, how much more or less will you buy or sell?' Business people always have to make guesses about what responses will happen as price changes are made. Price responsiveness (elasticity) is an interesting and useful concept because of the way it relates to total revenue of the business. Total revenue is the amount of the product sold multiplied by the price received. The law of demand holds that as the price of a product goes down people will buy more of it, and as the price of a product goes up people will buy less of it. So, as price changes, how much more or less will people buy, and therefore how much will total revenue change? If the proportionate fall in the price, which applies to all units sold, exceeds the proportionate increase in the number of units of output that are sold, total revenue will fall. For instance, a 10% decline in price of a good or service that induced only a 5% increase in sales would have the net effect of reducing total revenue received by the seller. Alternatively, a 10% increase in price that led to only a 5% fall in quantity demanded would lead to an increase in the total revenue received by the seller. If a small change in price induces a large change in quantity supplied or demanded, then the response is described as being highly responsive or elastic – that is, the percentage change in quantity exceeds the percentage change in price.

If a large change in price is accompanied by a small change in quantity demanded, then the response is said to be not very responsive, or inelastic. When faced with an inelastic demand, the seller could increase price and thus increase total revenue. The loss in revenue from the reduction in the quantity of sales will be outweighed by the increase in total revenue from the remaining sales – that is, selling less but at a higher price per unit.

When faced with an elastic demand, a seller could cut the price and increase total revenue. The loss in revenue per unit sold will be outweighed by the overall increase in revenue from the increase in total sales – that is, selling more at a lower price per unit. Business people cannot avoid making guesses about how responsive their customers are going to be to a change in price, and may do this explicitly or implicitly (even without ever having heard of the term 'elasticity'). Ultimately, the pricing decisions made by people running businesses are based on the expected response of customers and the expected effect on total revenue.

A further critical concept in economics is the notion of time period involved for any particular question being asked – commonly distinguished as the short run and the long run. In the short run a business cannot change the basic size of its operation by investing further in capital equipment, called fixed investment. In the short run a business can only increase output by working the existing fixed investment harder. This is done by applying more variable inputs, such as adding more casual labour to a processing line and running it faster or for longer. The long run is defined as the situation where a business can change the size of the whole operation by investing further capital and adding more equipment or new technology, and so on. This distinction between short and long run is important, because it has implications for the costs a firm has to recoup over different time periods.

It follows that the distinction between fixed and variable costs is useful when it comes to analysing the operations of a business. Fixed costs are those costs incurred whether the firm produces something or nothing. They are not avoidable, regardless of how much the firm produces. Fixed costs include depreciation of the capital equipment due to obsolescence, permanent labour, and land rates. Variable costs vary directly with the amount of product produced, and are avoided when nothing is produced. Variable costs are costs incurred to enable the fixed investments to be used to produce something, such as power, water, chemicals, raw material inputs, distribution and selling costs, commissions, casual labour, and marketing costs.

The distinction between fixed and variable costs is important, because these different types of costs have different implications for management and decisions. Fixed costs are not able to be avoided unless the business is closed down. This means that in the long run a business has to cover all its costs, both fixed and variable, and make a profit sufficient to give a return on capital comparable with alternative uses of the capital. Otherwise the capital will be eventually forced out of this business and into another investment. However, in the short run a business needs only to cover variable costs for it to remain in business. That is, if a business is facing a fall in the price of its product and this means that at current levels of output the receipts of the business won't cover all costs, what should it do? In the short run it makes sense to use more variable inputs as long as the revenue received per extra unit sold exceeds the extra variable cost associated with each unit sold. If so, each extra sale contributes something towards the fixed costs, even though not covering fixed costs fully, and therefore reduces the overall loss experienced in the short run. Over a longer period, however, all costs have to be covered. The short-run shut-down point is when the revenue from each unit of output sold becomes less than the variable cost of producing that unit of output. To produce in this situation means that variable costs are being lost as well as fixed costs. In this situation, total losses are minimised if the operation is shut down and only fixed costs are lost.

Another critical concept in management economics is the marginal effects of added inputs and extra output on profit. This is the idea of using a bit more of this input, a bit less of that input – what is the overall total effect?; or of producing a bit more of this output, a bit less of that output – what is the overall total effect? As more and more variable inputs to production are added to the fixed inputs, the extra output resulting from the extra variable inputs generally eventually declines. If this wasn't the case, then one factory (fixed input) could produce everything simply by adding more and more variable inputs to it. This isn't what happens. Instead, extra production resulting from extra variable inputs may initially increase at an increasing rate; then it may increase at a constant rate as more variable inputs are added; and then, with further variable inputs, output will increase at a decreasing rate. Then, if too much variable input is added to the fixed input, the effect eventually of extra variable input will be to cause a fall in total production. This phenomenon is known as an outcome of *the law of diminishing marginal returns to variable inputs*. The practical consequence of this effect is that as output increases, and more and more variable input is needed to make an extra unit of output, the cost of getting an extra unit of production from the activity increases. This is why producers will only supply extra output at higher prices – it costs more per unit to produce extra output.

The effect of increasing output on costs is ambiguous because, while variable costs per unit may increase because of the effect of the law of diminishing marginal returns, total fixed costs don't change. But the fixed costs per unit of output fall as more and more output is produced. Variable and fixed costs per unit of output are called average variable costs and fixed variable costs. The fixed costs of each unit of output decline as output increases and variable costs per unit of output increase. The net effect of increasing production is the impact on total cost per unit of output. Generally, as production increases from low levels the effect of spreading fixed costs and falling or constant variable costs per unit of output is to cause the total cost per unit of output to fall. Eventually the cost-reducing effects of spreading fixed costs over more output are outweighed by the cost-increasing effects of increasing variable costs and total cost per unit of extra output increases.

Revenue per unit of output minus total cost per unit of output equals profit per unit of output. This leads people to think that the most profitable amount of output for a firm to produce would be where the total cost per unit of output is at the lowest level for the particular operation. However, getting the maximum profit from total output isn't the same thing as getting maximum profit from each unit of output.

Maximum profit from total production involves getting any profit that can be had from each unit of output. Thus the important focus in business decisions isn't on the average total cost of each unit of output compared with the average revenue from each unit of output sold. Instead, the more useful way of thinking is: 'If I produce another unit of output, how much will it cost (how much will it add to total costs) and how much extra revenue will it bring in (add to total revenue)?'

The rule for maximising profit is: 'To decide on the quantity of output that will maximise profit, compare marginal cost with marginal revenue. The profit-maximising practice is to produce an extra unit of output as long as it adds more to total revenue than it adds to total cost. If this is the case, then a bit more profit is added to all the previous profit earned from previous units of production.' This process, producing a bit more output as long as extra revenue exceeds the extra cost of doing so, continues until the revenue from the last unit of output just exceeds the extra cost of the last unit of production. Then total profit is maximised (ignoring, for the moment, the reality that this theoretical level of output can never be attained because there are risks and uncertainties about precisely the cost and returns from extra output). The marginal way of thinking about profit maximising is extremely valuable in business decisions.

## Farming and the economy

Farming in Australia is a high-risk activity because production and prices are very volatile. The success or failure of agricultural activity in Australia is determined mostly by the nature of the season and the state of world demand and supply of agricultural commodities. Prices are determined by long-run and short-run changes in both supply and demand, both locally and overseas. The supply and demand of agricultural products are affected by economic developments and political decisions made locally and around the world to do with economic growth, access to markets, wars, economic slumps and political crises. Changes in economic activity in the major world economies translate into significant swings in the volume of trade and prices received for products. Less obviously, but equally importantly, the success of farming is also intricately affected by and linked to what happens in the non-agricultural sectors of the domestic and international economies.

Agricultural activities contribute to the process of growth of an economy and the wellbeing of people in five main ways. These are:

- increasing production of food and fibre to provide for an increasing population;
- releasing a surplus of labour for other business activities;
- generating net income to provide capital for investment in, and further development of, primary, secondary and tertiary industries;
- providing a market for the output of the other business activities; and
- earning export revenues, which enables the purchase of imports.

The importance of these contributions varies from country to country, depending on the stage of development and the endowments of resources of countries. From the beginning of European settlement in Australia, a particularly important role of agriculture has been to earn income from exports. Agricultural export earnings make it possible to buy raw materials, capital and consumer goods from businesses in other countries.

Key indicators of the importance of agriculture in an economy are the share of national income due to agriculture, the proportion of total exports, and the proportion of the labour force employed in agriculture. In each of these measures, agriculture in Australia has declined in importance relative to other sectors of the economy. Agriculture's contribution to Australia's national income declined from around 40% in 1840 to 20% in 1900, 14% in 1960, 4% in 1990 and to less than 2% in 2005. Rural production contributed 85% of total exports in 1950 and about 25% in 2005. Agricultural export earnings contribute around 15% of national income.

A number of features of the economy beyond the farm gate dictate (a) that the agricultural sector's share of the gross national income will decline, and (b) that incomes per farmer will inevitably decline relative to everyone else in the economy unless some farmers get out of farming. These two features are so well-grounded in theoretical analysis and practical observation that they can be regarded as 'rules' of the game. As economic growth occurs and people become wealthier, their requirements for food and fibre tend to increase at about the rate of population growth, which also slows with increasing development. With increasing wealth, people spend less of their increased income on the food and fibre products of farming businesses. There is a limit to how much of these commodities people can use, no matter how wealthy they become. At the same time, people spend their increasing wealth on non-agricultural goods and services. The result is that with economic growth and increases in the gross national income, the percentage share of that increase that goes to the farm sector declines relative to the share of income that goes to the rest of the economy. In wealthy countries whose national income is growing, the proportion of extra income that is spent on agricultural commodities is around 10–20%. In poorer countries this can be around 60–70%, or even more. The consequence of the rest of the economy expanding in line with people's demands, and agriculture declining as a percentage of total economic activity, is that unless there were fewer farmers over time the income going to each farmer would decline relative to the incomes going to people in the continually expanding non-farm sector of the economy.

A feature of the supply of agricultural commodities is that supply tends to change only slowly in response to changes in prices received and costs. This is because many farmers' costs are unavoidable (fixed). Many costs are incurred whether farmers produce or not. So, as long as the price farmers receive for their production more than covers the avoidable (variable) costs, such as the seed, fertiliser and fuel, they may as well keep producing, for the short run at least. Relatively large falls in commodity prices have to occur before producers stop producing. Often farmers' land, labour and capital are still in their best uses, despite pronounced falls in returns to those factors. Also, most farmers are doing what they are best at, and what they love doing. So, even when prices and incomes decline considerably, there will tend not to be a correspondingly large fall in total supply of farm products.

Australian farmers have been fighting a cost-price squeeze for decades – the real (no inflation) prices of the things they buy rise faster than the real prices of the products they sell. Farm costs rise because farm businesses have to increase productivity and require more and improved inputs. Farm costs also rise because farmers have to compete with the rest of the economy to obtain the resources of land, labour, management and capital that they need. At the same time, the demand for farm production isn't highly responsive to reductions in price. People don't want lots more food and fibre, even if the price falls a lot. Large changes in price are required for demand to adjust to even small changes in supply. Also, the amount that can be sold, and the prices received, can vary greatly and rapidly. Access to markets is often restricted and supply fluctuates. These effects, along with rapid increases in production capacity, mean that the prices farmers receive for their products decline over time in real terms. Declining prices mean that farmers have to improve their productivity continually so as to maintain their income. Most farmers do so. To maintain their income, farmers become more productive largely by increasing costs and adopting further supply-increasing technology, and expanding the size of their operations. This phenomenon, of farmers continually having to run faster to keep the effects of the cost-price squeeze at bay, to avoid persistently receiving lower incomes than are received on average in the rest of the economy, is known worldwide as 'the farm problem'. Despite this phenomenon, farm businesses operate profitably by continually changing how their businesses operate and increasing output per unit of input, called productivity.

In large part the story of Australian farming is the story of farmers adapting to changes in economic, technical and natural circumstance in order to survive and succeed. Farm businesses that are better endowed with resources, especially land, capital and management skills, and who master the increasing information needs and complexity of their farming systems, and have reasonable luck, can and do thrive, while others steadily leave the land. The least efficient producers, those with the relatively higher cost structures (often the small producers with consequently high overhead costs per unit of output), are put under competitive pressure by the cost-price squeeze. These farmers either adjust their operation, or become reconciled to receiving low incomes for a long time, or extract assistance through the political process, or move out of the industry. Adjustments (some get bigger, some diversify, some get out), plus large gains in productivity, largely averted the persistent, widespread low-income problems that have beset the farming sectors of other developed economies and that are dealt with in the United States and Western Europe by massive handouts to farmers from taxpayers and consumers. While small pockets of low-income farmers can always be found in any industry in Australia at any time, this phenomenon usually has more to do with temporary market slumps, bad investment and management decisions, and personal preferences, than with long-term structural problems in agriculture.

Farm businesses get into financial difficulties because the resources they command are, or their command of resources is, inadequate and they cannot generate sufficient cash surplus to service their debts and meet family living requirements. When relative prosperity and optimism prevail in an industry or area, farmers in severe financial difficulties are able, if they wish, to leave farming by selling out at a reasonable price to farmers who are expanding their businesses. Farmers leave farming, but generally their own resources and resourcefulness let them dictate some of the terms of their leaving the land. While they might not get their initial price, they generally get out with a bankroll that by community standards isn't too bad. This process of some farmers getting out and some getting bigger is called adjustment. It is an inevitable and continuing process in agriculture in modern mixed economies. Indeed, such is the inevitability of this process that governments of all political persuasions actively assist such adjustment when economic recession takes over in rural areas and the adjustment process falters or stalls. Governments also provide special welfare aid for farm families who are suffering hardship, for any reason.

Falling output prices and rising input costs don't necessarily cause farm incomes to fall. Gains in output per unit of input, called productivity, enable some farmers, but not all, to maintain or increase their incomes. In Australia, a number of factors have enabled many farmers to increase their productivity and maintain average levels of total income comparable to incomes received in the other sectors of the economy. These include the large land base, the substitution of capital and new technology in place of labour, the widespread existence of mixed farming with a range of alternative combinations of activities, and government-sponsored research and extension.

The riskiness and uncertainty of production is one of the most notable features of farming in Australia as compared with most of the rest of the world's agriculture. Seasonal conditions are so variable that the supply of farm production can range from a lot to a little over a short time. As well as coping with volatility in seasonal conditions, Australian farmers must sell large proportions of their output at unstable world prices. On world markets there are large shifts of supply (especially from seasonal conditions) and in demand (from changes in fashions, tastes and income) that destabilise price. Furthermore, for many products world trade is only a small proportion of output, which makes for price instability. Export markets are relatively small and are residual markets. Only a small proportion of total agricultural production is traded on international markets. Relatively small increases in production cause relatively large falls in prices received. For example, of the 600 million tonnes of wheat produced throughout the world each year, only about 100 million tonnes is sold on international markets. If 20% of total wheat production entered world markets, then a 10% increase in total production represents close to a 50% change in the supply for trade on world markets. Given the nature of the demand, this causes a large drop in prices received for wheat on the world markets. Increasingly the 'going price' on overseas markets depends considerably upon the access that sellers are granted to those markets (through the plethora of bilateral and multilateral trade agreements) and on the extent to which competing sellers are disposing of subsidised production.

## Agribusiness

Agriculture makes agribusiness different from business in general. Businesses that deal with agricultural goods and services and agriculture-related goods and services operate under different circumstances from other businesses in general that don't have an agricultural connection. Understanding the operation of businesses that deal with agricultural goods and services is enhanced markedly by an understanding of the agricultural aspects of the production of such goods and services. This is because the nature of agriculture – the biology, the markets, the seasons, the risks and uncertainties – impose a special set of conditions and requirements on the management of businesses related to agriculture.

When pondering whether a business is an agricultural-related business, a useful test is to consider how influential the things that happen on farms are. Conceptually, the more directly a business is related to what happens in the farm business, the more it is in the area known as agribusiness. The agriculture that impacts on that business creates a need to modify the usual principles and analytical approaches applying to business in general.

Conceptually, when agribusiness is looked at in terms of outputs, the boundaries of agribusiness and business in general blur and merge the more the product is transformed from raw agricultural material to something else. Agribusiness activity becomes more clearly defined the closer we move towards the farm sphere and the less the product has been transformed. The more the product is transformed, the more likely it is that activities in these areas are already covered by the traditional areas of study called economics, business management and business marketing. Is the pie seller at a football match engaged in agribusiness, or is it simply business?

Agribusiness is an interdisciplinary area of study. What particular disciplines are brought to bear on a question to do with agribusiness depends on the question that is being asked. The key to analysing the management of a business and to problem solving in the real world is to bring to bear on the question at hand the appropriate mix or balance of disciplinary knowledge – and the appropriate balance depends on the question. The 'optimal degree of generality' refers to bringing the right mix of disciplinary knowledge to bear on the question at hand in order to solve problems in a sensible, whole, manner. It follows that study of agriculture-related production of goods and services should cover well the science, economics and human behavioural disciplines, because various mixes of knowledge from these areas are needed to solve different problems in agriculture-related production. The interdisciplinary field called agribusiness derives from the theoretical foundations of the disciplinary fields of agricultural economics, agricultural science and human behaviour, and the interdisciplinary fields of business management and business marketing. The 'whole business' approach can be brought to the whole marketing chain. Technical aspects of the nature of the product and the processes involved in production go a long way to explaining why, for instance, meat, wheat, wool and dairy processors operate and market the way they do, individually and in aggregate.

The environment in which Australian agribusiness firms operate has changed markedly since the 1980s. Deregulation of the exchange rate, financial sector, labour markets and trade arrangements has led to greater exposure to competition in all markets. A major consequence of increasing deregulation of economies is greater uncertainty. At the same time, more avenues to manage volatility become available. Furthermore, changes in the level of economic activity and in key economic variables such as rates of foreign currency exchange, or interest and inflation, have a more immediate and direct impact on the welfare of individual farmers than was the case in the more regulated business environment of the past. International economic developments also directly affect individual economies to a greater extent than in the past. For instance, flows of capital are more readily 'internationalised' than ever before. This has a number of implications. Interest rates, once largely under the influence of monetary and fiscal policy, are now also very much the product of trade policy. Nowadays, the effects of changes in the economies of the major developed countries, as well as in the lesser developed but developing countries, are likely to translate more readily into effects on Australian businesses, including farmers. In agricultural business in Australia, like all business, there are good-, medium- and poor-quality managers. In the future, as in the past, those people involved in agribusiness who succeed will be those whose businesses are appropriately capitalised and who best manage the risks and uncertainties – that is, those managers who are intellectually best-equipped to master information and form sound judgments.

The next chapter is about ways of thinking about analysing farm businesses.

## CHAPTER SUMMARY

*Economics is the core discipline of farm management analysis. The whole farm approach is how the economic way of thinking is applied in farm management analysis.*

## REVIEW QUESTIONS

1 What is the basis of the case that economics is the core discipline of farm management analysis?
2 People sometimes talk about 'the whole farm approach to farm management economics' as though there is some other approach. There is no other approach. The whole farm approach is the economic approach. Explain the whole farm approach.
3 'It is better to be roughly right than precisely wrong.' What does this statement mean, in the context of farm management analysis?
4 Risk is the outstanding characteristic of Australian agriculture. In what main ways does risk affect farming?
5 'Start with the farm family' is the rule about farm management analysis. Why?
6 Returns to agricultural investment can be commensurate with other investments in the economy – for the best-run agricultural businesses. Discuss.

## FURTHER READING

Boulding, K.E. 1956, 'General systems theory: The skeleton of science', *Management Science*, vol. 2, pp. 197–208.

Boulding, K.E. 1974, 'General systems as a point of view', in L.D. Singell (ed.), *Collected Papers Vol. 4, Toward a General Social Science*, Colorado Associated University Press, Boulder, Colorado.

Brennan, L.E. and McCown, R.L. 2002, 'Back to the future – reinventing farm management economics in farming systems research', paper presented to the 46th Annual Conference of the Australian Agricultural and Resource Economics Society, Canberra.

Brennan, L. and McCown, R. 2003, 'Making farm management research relevant to farm management practice', invited paper presented to the 47th Annual Conference of the Australian Agricultural and Resource Economics Society, Fremantle.

Collins, J. 2001, *Good to Great: Why some companies make the leap . . . and others don't*, Random House, Sydney.

Dillon, J. 1965, 'Farm management as an academic discipline in Australia', *Review of Marketing and Agricultural Economics*, vol. 33, pp. 175–89.

Dillon, J.L. 1976, 'The economics of systems research', *Agricultural Systems*, vol. 1, pp. 1–10.

Dillon, J.L. 1979, 'An evaluation of the state of affairs in farm management', *South African Journal of Agricultural Economics*, vol. 1, pp. 7–13.

Dillon, J.L. 1980, 'The definition of farm management', *Journal of Agricultural Economics*, vol. 31, pp. 257–8.

Kingwell, R. 2002, 'Issues for farm management in the 21st century: A view from the West', *Australasian Agribusiness Review*, vol. 10, paper no. 6, www.agrifood.info.

McConnell, D.J. and Dillon, J.L. 1997, *Farm Management for Asia: A systems approach*, Food and Agriculture Organization of the United Nations, Rome.

McCown, R.L. 2001, 'Learning to bridge the gap between science-based decision support and the practice of farming: Evolution in paradigms of model-based research and intervention from design to dialogue', *Australian Journal of Agricultural Research*, vol. 52, pp. 549–71.

McCown, R.L. 2002, 'Changing systems for supporting farmers' decisions: Problems, paradigms and prospects', *Agricultural Systems*, vol. 74, pp. 179–220.

McCown, R.L. 2002, 'Locating agricultural decision support systems in the troubled past and socio-technical complexity of models for management', *Agricultural Systems*, vol. 74, pp. 11–25.

McCown, R.L., Hochman, Z. and Carberry, P.S. 2002, 'Probing the enigma of the decision support system for farmers: Learning from experience and from theory', *Agricultural Systems*, vol. 74, pp. 1–10.

McGregor, M., Rola-Rubzen, F., Murray-Prior, R., Dymond, J. and Bent, M. 2003, 'Farm management – bugger the roots, where is the future?', paper presented to the 47th Annual Conference of the Australian Agricultural and Resource Economics Society, Fremantle.

Malcolm, L.R. 1990, 'Fifty years of farm management in Australia: Survey and review', *Review of Marketing and Agricultural Economics*, vol. 58, pp. 24–55.

Martin, S. and Woodford, K. 2003, 'The farm management profession in New Zealand: Where are our roots?', invited paper presented to the 47th Annual Conference of the Australian Agricultural and Resource Economics Society, Fremantle.

Mullen, J.D. 2002, 'Farm management in the 21st century', *Australasian Agribusiness Review*, vol. 10, paper no. 5, www.agrifood.info.

Ronan, G. 2002, 'Delving and divining for Australian farm management agenda: 1970–2010', *Australasian Agribusiness Review*, vol. 10, paper no. 7, www.agrifood.info.

Schultz, T.W. 1939, 'Theory of the firm and farm management research', *American Journal of Agricultural Economics*, vol. 21, pp. 570–86.

3

# Analysing a farm business

The focus of farm business analysis is the current situation, the potential and constraints, and the risks and uncertainties. The principles that need to be applied to understand the mix of economic, technical, financial, growth and risk factors that are part of the whole farm system, and the environment in which it operates, are explained.

## TECHNICAL BASIS OF ECONOMIC, FINANCIAL, GROWTH AND RISK ANALYSIS

Agricultural production is the process of using resources to make goods or provide services. Producers can use combinations of the three broad classes of factors of production called labour, capital and raw materials to produce one or many products. A key element in a farmer's decision about what to produce and how to produce it is the objective of getting more, or even the most, out of the limited amount of resources with which he or she has to work.

There are three basic technical relationships in production:

- the relationship between the amount of a resource used and the amount of production (called input–output);
- the different ways resources can combine and substitute for one another in the production process (called input–input); and
- the relationship between different products that can be produced with the resources that are available (called output–output).

## Input–output in the current production period

Consider a simple hypothetical situation: a farmer has some land, labour, machinery and equipment, and the finance to buy the seed, fertiliser, fuel and chemicals needed to work the land and grow a crop, using his or her own labour. The farmer has land, labour and capital. The focus of economic thinking is on using these resources efficiently. This means using resources in such a way that the user gets the most of what he or she wants out of them. The time period of interest is important. In the immediate production period, called the short term or short run, only the inputs to production that can be changed and that can affect output in this time are of interest – for example, fertiliser, water, chemicals, seed and fodder. These inputs are said to be changeable or 'variable in the short run'. The other inputs of land, labour, machinery and livestock cannot be changed in the immediate production period and are said to be 'fixed in the short run'.

The relation between resources used in production (called inputs – in this case, labour, land and capital) and the production which results (called output – here, it is the amount of crop harvested) is called the production function or response function. A general principle of production is that where there is a fixed amount of one resource (for example, land), more output can be obtained only by adding other resources to it, such as labour, fertiliser, fuel, machinery services, and so on. In this case, land is the fixed input. It is 'fixed' because once the crop is planted the farmer cannot, in the short run, change the amount of land being used. The land is used for the crop that has been planted regardless of whether a large or small yield is harvested. Other productive resource inputs such as labour, fertiliser, chemicals or seed can be added to the production process in varying amounts.

Consider the relation that exists between the total amount of fertiliser applied to the crop throughout the whole production process, and the total

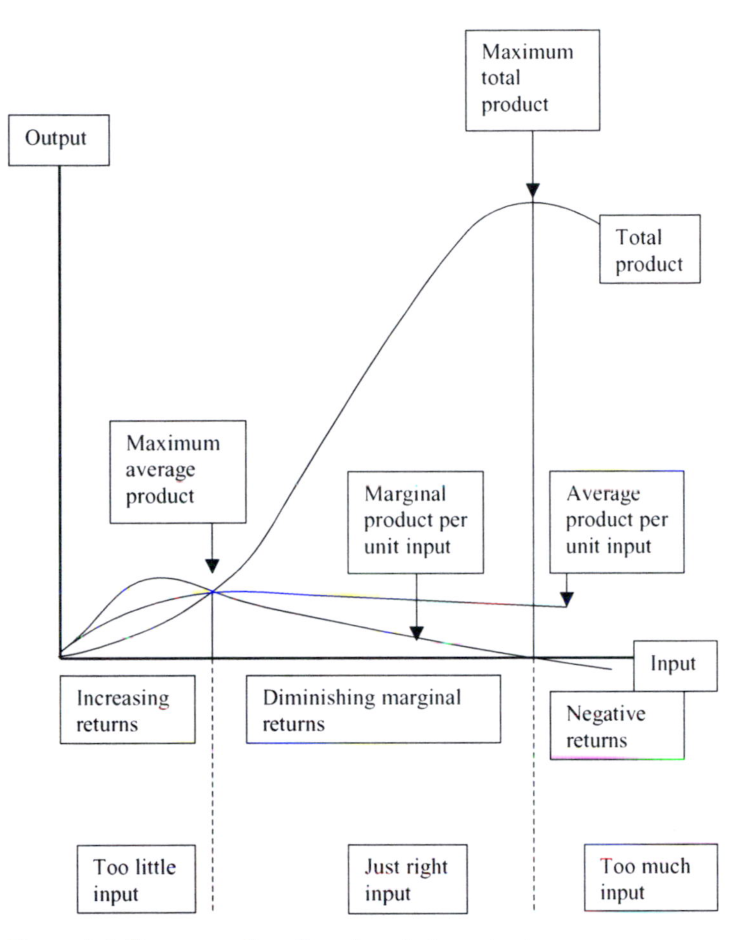


**Figure 3.1** Response function: Inputs to output

amount of crop harvested. With all other requirements of the crop at an adequate, constant level, a relationship can be shown to exist between total fertiliser used and total yield of crop. The relation between the quantity of fertiliser and the amount of yield shows the response of yield to fertiliser, as demonstrated by the response function shown in Figure 3.1.

The extra production that comes from an extra unit of input is called the extra output or the marginal product. Without any fertiliser there would probably be very little crop output. At a low amount of the variable input (fertiliser), each extra unit of fertiliser will contribute more to total output than each previous application of fertiliser. For example, application of 80 kilograms of fertiliser may be more than twice as useful as one application of 40 kilograms of fertiliser. Suppose the first three applications of fertiliser, will each cause quite a significant yield response. That is, for the first three applications, each of 40 kilograms of fertiliser, each extra 40 kilograms of fertiliser contributes more to total output than the previous 40 kilograms of fertiliser. This situation is called increasing returns, and is shown in Figure 3.1. As fertiliser input increases, a level will be reached where each extra unit of fertiliser still adds to total crop production, but each increase is less than the contribution of the previous application of fertiliser. This phenomenon is called diminishing marginal returns and is shown in Figure 3.1. Extra output is positive but getting smaller with each extra input of fertiliser. Ultimately, an extra application of a unit of fertiliser won't add anything to total crop production, and more fertiliser will cause total output to fall. This could be due to the level of total fertiliser being so high as to be toxic to some of the plants. The extra product resulting from each extra input is negative at this level of input use. This is known as the stage of negative returns to extra input, as shown in Figure 3.1.

This technical relationship between variable inputs and outputs, where the extra output from extra inputs becomes less and less, is called the law of diminishing marginal returns. This 'law' describes the practical outcome where if increasing amounts of one input are added to a production process, while the use of all other factors is held constant, then the amount of extra output resulting from each unit of increased input will eventually decrease. Knowing the nature of this response of output to variable inputs, and without knowing anything about the cost of the variable input or the value of the output, it can be seen that some levels of input would be sensible, and some levels of input would be either too high or too low. For instance, regardless of how much fertiliser might cost and what the crop output is expected to be worth, it isn't sensible to apply so much fertiliser to the crop that the extra fertiliser reduces total crop production below what could be produced with less fertiliser. In Figure 3.1, up to the level of B kilograms of fertiliser applied per hectare, the total output from the land increases with each extra application of kilograms of fertiliser. Beyond this, total output decreases.

As well as looking at total output to decide how much fertiliser to use, it is instructive to look at the amount of output resulting from each of the inputs used. This is called the average output or average product (Figure 3.1). The average output of an input is the total amount of output divided by the total quantity of that particular input that has been used to produce that total output. In this example the average output per extra unit of fertiliser initially increases as the amount of fertiliser added increases. It makes good sense to use inputs at least up to where the average output of units of fertiliser input is highest.

In the example of fertiliser and land inputs to produce a crop, as more fertiliser (variable) input is applied to the land (fixed) input, total output increases rapidly. When the extra output is increasing (adding more and more to the total produced), this is pulling up the average output of all the units of fertiliser applied previously – for example, in Figure 3.1 the average output of an input is highest where the marginal (extra) output or marginal product equals it. At low levels of input, marginal product is greater than average product, and with further applications of fertiliser the extra output raises the average output. After the point is reached where marginal product equals average product, then the extra output from extra fertiliser is less than average output and the extra fertiliser is pulling down the average output of all previous fertiliser that has been applied.

The practical rule that comes out of this information is that it makes good sense to use a variable input such as fertiliser at least up to the level where the average production from each application of fertiliser is highest. There is no sense applying less than this amount of fertiliser, because up to this point application of an extra kilogram of fertiliser will increase the average output of all the previous fertiliser applied.

Technically sound levels of fertiliser have been identified – that is, between the amount of fertiliser that gives the highest average output per unit of fertiliser used, and the total amount of fertiliser that gives most total output. Put another way, two 'zones of input use' have been defined that don't make technical sense: a zone of 'too little' fertiliser and a zone of 'too much' fertiliser (Figure 3.1). As well, the range of input levels that are technically sound has been defined ('just right', Figure 3.1). The important point is that the economically 'best' level of input use will be somewhere within the technically sensible range of fertiliser application – that is, in the 'just right' zone. The precise amount of fertiliser to use to make the most profit depends on the cost of the fertiliser that is applied and the value of the output.

In practice, response functions on particular paddocks are never known with precision, though farmers have a good idea, within a range, of the output responses they can expect from extra variable inputs, if the season turns out as expected. As well, there is no certainty about actually achieving the output responses to the fertiliser that are expected. The expected result might not happen, for one or more of three main reasons: (a) the conditions under which the crop is grown will differ from the conditions in which the response function was found to operate; (b) the season may differ from the expected season; and

(c) prices may not be as expected. However, the principles of diminishing marginal returns and the marginal way of thinking that lead to conclusions about what to produce and how to do it on the basis of the estimated technical relations between inputs and outputs, and extra returns and extra costs, are the foundations of good farm management analysis.

## Input combinations in the current production period

Suppose that two variable inputs are used in production – fertiliser and quantity of water applied. There are a number of ways in which these two inputs could be combined to produce a given amount of crop from the land. The farmer could combine plenty of water with little fertiliser, or a lot of fertiliser with less water, or use some combination between these extremes. Remembering the law of diminishing returns, combinations of a lot of both inputs ('too much input' zone) could mean that the extra product of one or both inputs is negative. Such combinations wouldn't be sensible, in technical or economic terms. Combinations of too little of either input, with average product per input less than maximum ('too little input' zone), wouldn't be technically or economically sensible either.

The rate at which one input can replace another input in the production process, while maintaining a certain level of output, is of special interest. If a particular quantity of the fertiliser is applied to a crop, how much water needs to be applied to maintain a specified level of production? If less water is applied, how much extra fertiliser is needed? At any given level of output, inputs may substitute for each other at an increasing rate (one input replaced by less than one substitute input), a constant rate (one input for another input), or more commonly, at a diminishing rate (more and more of one input required to replace a unit of the other input). Again, the input combinations where the extra product of each unit is positive are technically sound. Here, in technical terms, inputs are not being wasted (similar to the 'just right' zone in the single input case). These ideas hold for any number of resources used in production. The economic principle is to use combinations of inputs that cost the least to produce a given amount of output.

## Product combinations in the current production period

The third case is where combinations of resources can be used to make more than one product. The choices are to make either one product or the other, or some combination of the two. For example, there are a number of ways that producing

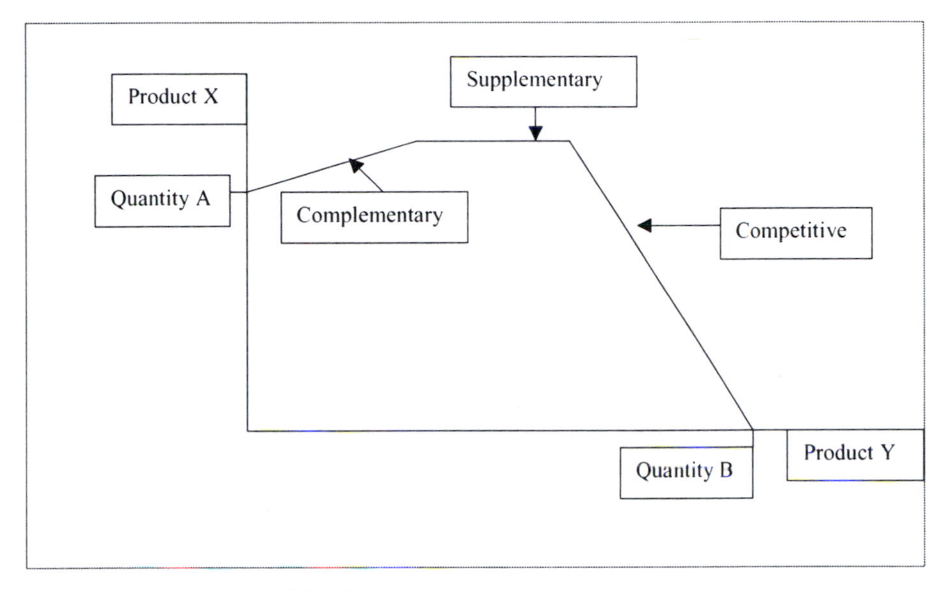


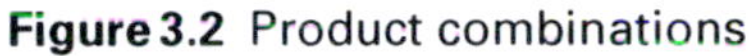
**Figure 3.2** Product combinations

product X might affect the production of product Y (Figure 3.2). The two products could help each other in production, meaning that if more X is produced it will contribute to more of Y being produced – that is, X and Y are complementary. Or X and Y might be supplementary, meaning that more of one doesn't affect the amount of the other that is produced. Products X and Y could be competitive for the farmer's land, labour and capital. (With more of one product, less of the other can be produced.) If a farmer wanted to make the best use of the resources available, on technical grounds it would pay to make use of any complementary or supplementary effects between products. The economic objective is to produce combinations of output that give the most profit.

So far, it has been shown that:

- it is worthwhile to use an extra unit of input if it will raise the average output of the previous units of the input used; and
- it is time to stop using more of the input if a bit more of it will reduce the total product.

These principles are useful ways of thinking about production decisions in farming. This 'cause-and-effect' understanding of the technical relations between amounts of inputs and amounts of output (and between different inputs and different products) provides a basis on which to make decisions about what to produce and how to do it. How can these technical relations be known? There are two important sources: the farmer and the researcher. Farmers have a reasonably good idea of the size of the effects that different amounts of input will have on output on their farm. Researchers can discover the relationship between inputs and outputs under experimental conditions. Their results, if interpreted cautiously to allow for the differences between experimental conditions and the real farming situation where farm yields might be two-thirds of research experimental yields, can form the basis of sound farm management analysis.

## ECONOMIC AND FINANCIAL CONDITIONS

The philosophy of farm management economic analysis is that it is more useful to solve the whole of a problem in an approximate manner than it is to solve a small part of a problem extremely well, and at the same time leaving out significant parts of the question at hand. Thus economic analysis is an interdisciplinary activity: the technical basis of agricultural economic analysis has to be sound, with the human element included, the economic and financial aspects enumerated, the risk elements considered, and the institutional factors acknowledged. The aim of agricultural economic analysis is to put the components of a system together into a whole, albeit a simplified whole, but one in which all the elements that have an important bearing on the question are considered. By contrast, a scientist focuses in depth on part of this whole, and while dealing completely with and 'fixing' the small part of the whole, will have difficulty explaining how the whole works or prescribing solutions to problems of the whole.

### Economic principles of using resources for agricultural production

Economic principles can be used in deciding what to produce and how to do it efficiently. If a producer chooses to use resources in a certain way, then they have given up the opportunity to use their resources in some alternative way. If a farmer does one thing, what is he or she giving up the chance to do? Which choice makes the farmer better off? Producers need some way of comparing alternative uses of resources. Comparisons require common units of measurement. Dollars are used to measure as many of the costs and benefits as can be valued in this way. Prices have to be taken into account when making choices about what to produce and how to produce it. The costs of inputs and prices of outputs are used to calculate how much production, and what combinations of inputs and outputs, are most profitable. This information can be used to work out whether a farming activity, or a change in the way the farm is operated, is going to make the producer better off in some way. The 'better off' in this case is 'profit'. By 'profit' is meant a surplus of income from production after all of the costs incurred in production have been deducted.

## Input–output in the current production period

Where there was only one variable input, fertiliser, it was shown that the extra output from adding extra fertiliser at first increases, then declines. Further, the best level of fertiliser application was somewhere in the 'just right' zone (see Figure 3.1). Knowing the cost of each extra application of fertiliser, how much extra output will be produced by each extra application of fertiliser, and what that output is worth (how much it could sell for, or what it would cost to buy), it is then possible to work out what level of fertiliser will contribute most to total profit. This is done for the fixed level of all the other costs involved in running the business. With more fertiliser, where extra outputs are still positive but becoming smaller, the cost of extra units of fertiliser may eventually exceed the value of the extra output that results. If so, further fertiliser isn't worth applying. It would reduce the total gains that have been made from all the previous applications of fertiliser.

To maximise profit from adding an input such as fertiliser to production, the decision rule, providing you can finance it, is to fertilise the land until the extra cost of fertilising almost equals the extra return from doing so – that is, up to the number of kilograms of fertiliser where extra revenue almost equals extra cost. Too little fertiliser will mean some more profits could be made by adding more fertiliser, and too much fertiliser reduces the profits that are made. If it can be financed, the rule is: to maximise profit use inputs to the level where the extra return that results equals the extra cost of the extra fertiliser used. The profit-maximising principle – that extra return should equal extra cost – indicates the best amounts of inputs to use. If the farmer cannot afford this profit-maximising level of inputs, then the decision rule is to use as much input (in this case, apply as much fertiliser) as can be afforded.

## Input combinations in the current production period

In the case where there is more than one variable input (for example, fertiliser and water), the most profitable combination of inputs depends on the cost of each input and the relationships between them. The aim is to combine these two inputs so as to produce output as cheaply as possible. From the costs of the two inputs, the total cost of different combinations of the inputs can be calculated and the cheapest combination(s) found.

When the price of an input changes, so does the cheapest combination to produce a given amount of output. For example, if the price of fertiliser rises relative to the price of water, then to maintain profits the farmer would produce the same output using less fertiliser and more water. Suppose that a large amount of one input is being used and the extra contribution to output from using an extra unit of input is becoming 'low' (because of diminishing returns). Suppose, also, that little of another input is being used and the extra return from using an extra unit of this input is still relatively high. Then, in order to produce the same output more cheaply, it is sensible to use less of the input with relatively low extra returns and more of the input with relatively high extra returns. This is the principle of substitution, or equi-marginal (equi-extra) returns, and is the key to maximising profit using a number of variable inputs by producing given amounts of output at least cost.

## Product combinations in the current production period

What is the profit-making rule when producing two or more products? If the products are complementary or supplementary, then it is good technical sense to produce both products at least up to where they start to compete for the limited resources. As shown in Figure 3.2, these two products are competitive when more *Y* can be produced only by producing less of *X*. The most profitable combination is the one that brings in the most income. To find the combination that brings in the most income, look at the technically possible combinations and the price of each product. Then calculate the total revenue each combination will provide. At the combination of two products which brings in the most income, the extra return from using available resources to produce more of one product is about the same as the extra return from using the resources to produce more of the alternative product. Here, there is no scope for further gains by substituting more of one product for less of another. This is an application of the principle of equi-extra returns.

To sum up: Applying the economic way of thinking about production in the immediate production period involves identifying the following:

- Input levels where marginal product is greater than average product, and where using more inputs means average product is being 'pulled up'. This is the zone of production called increasing returns and isn't rational because too little of the input is being used.

- Input levels where marginal product is less than average product, and where using more inputs means average product is being 'pulled down'.
- Input levels where total product is declining and marginal product is negative. This is an irrational zone of production because too much of the input is being used.
- Input levels where marginal product is less than average product but greater than zero. This is the rational zone of production, on the basis of physical responses to inputs. Within the range of inputs in this zone, the most profitable level of input use will be found.

The economic way of thinking about using resources leads to the conclusion that if there were plenty of funds, the most profitable amount of an input to use would be where the extra return just exceeds the extra cost of each extra input to the various forms of production carried out on a farm. However, there are never 'plenty of funds' available, relative to all the uses which could be made of them. More likely, funds would not be available to apply the profit-maximising amount of input to each activity. The aim, then, is to make the most use of the limited funds that are available for spending on production. This is done by first applying inputs to the activity where the extra return is highest, and then to the activity that has the next highest return, and so on. The most profit comes from applying the limited inputs to the alternative production uses until the marginal unit applied to each activity brings in roughly the same return. The same thinking applies to investing on and off the farm. If further investment of capital off the farm will earn 10% p.a. on capital, and further investment on the farm will earn 15% p.a. on capital, then further investment on the farm is warranted.

## Increasing farm profit in the current production period

A farm business uses resources to make products to sell for more income than the product cost to produce, and thus earn profits. The profit earned in a given period, usually a year, is estimated by matching all costs incurred over the period with the value of all output produced over the period. Even production that has not yet been sold but has been produced has to be counted in estimating the income and costs of a defined period. The main source of income in a farm business is from sales of crops and animals produced as part of the annual farming operation;

or from non-cash income, resulting from extra stocks on hand at the end of the trading year from production that has not yet been sold, called inventory change.

There are two broad types of farm costs:

- *Variable costs (sometimes called direct costs):* These vary directly as the output of an activity varies.
- *Overhead or fixed costs:* These are costs which don't change as small changes in the level of an activity are made.

Typical examples of variable costs are:

- chemicals and fertiliser;
- insurance on animals, crops and seed;
- irrigation running costs;
- fuel and oil;
- casual labour and shearing costs;
- marketing costs;
- bought feed;
- veterinary expenses; and
- repairs to machinery.

Some typical overhead costs are:

- permanent workers' wages and employee on-costs;
- annual repairs to water supply, roads, buildings and structures;
- insurance on employees, fixed structures and plant;
- telephone and business expenses;
- power costs;
- shire rates and land taxes;
- annual fixed water rates;
- depreciation of plant and improvements;
- consultants' and accountants' fees;
- operator's allowance for labour and management; and
- running cost of farm vehicles (not attributable to particular activities).

Profit is the surplus of income remaining after all the costs that were involved in earning the income have been deducted. For example:

Total income
*Minus* Variable costs
*Equals* Total gross margin
*Minus* Overhead costs
*Equals* Profit (usually called operating profit, or earnings before interest and tax (EBIT))

The main principles about economic use of inputs to produce output can be demonstrated within the following framework of gross income (GI), total variable costs (TVC), total overhead (or fixed) costs (OH), total gross margin (TGM) and operating profit (OP) – that is:

$$GI - TVC = TGM - OH = OP$$

Data on production, costs and income, relevant to a farm business, is given in Table 3.1 and Figure 3.3. The relationship between the main measures of cost is shown and the profit-maximising levels of input use are defined.

> Consider the information in Table 3.1 and Figure 3.3. In this example, the focus is on maximising profit in the short run. If the price received for wool is the same whether the farmer produces wool from 4,000 sheep or 9,000 sheep, and the overhead costs stay the same, what is the best number of sheep to run? The extra 1,000 sheep (from 8,000 to 9,000) add $30,000 to total variable costs and 4,400 kilograms to total wool. The extra return is only $5 per kilogram. Total profit would be reduced by running 9,000 sheep. The reason is that the marginal costs of the extra 1,000 sheep are high, because variable costs increase significantly at this stocking rate. This is because higher fertiliser costs are incurred to grow the required extra feed, extra casual labour is required, and the need for supplementary feed for all the sheep increases at the higher stocking rate. The most profitable stocking rate would be about 8,500 sheep (see Table 3.1 and Figure 3.3). At 8,500 sheep, the marginal cost per kilogram of wool is $4.50 and the most likely return is $5 per kilogram. This is near enough to the theoretical profit-maximising stocking rate ($MR = MC$), once it is remembered that there is some uncertainty about this estimate of marginal cost per kilogram, and there is a reasonable chance that in some years the costs will be higher because of a drought, or the price will be lower.

**Table 3.1** Production, costs and profit for a farm for existing and changed stocking rates

| Current situation | | Changed stocking rate | | |
|---|---|---|---|---|
| Number of sheep (wethers) | 4,000 | 6,000 | 8,000 | 9,000 |
| Wool cut/head (kg) | 5 | 5 | 4.85 | 4.80 |
| Total wool clip (kg) | 20,000 | 30,000 | 38,800 | 43,200 |
| Wool price/kg ($/kg net) | 5 | 5 | 5 | 5 |
| Gross income | 100,000 | 150,000 | 194,000 | 216,000 |
| Total variable costs | 36,000 | 60,000 | 96,000 | 126,000 |
| Total gross margin | 64,000 | 90,000 | 98,000 | 90,000 |
| Marginal cost/kg (as DSE and wool kg change) | | 2.4 | 4.1 | 6.8 |
| Overhead costs | 50,000 | 50,000 | 50,000 | 50,000 |
| Operating profit | 14,000 | 40,000 | 48,000 | 40,000 |

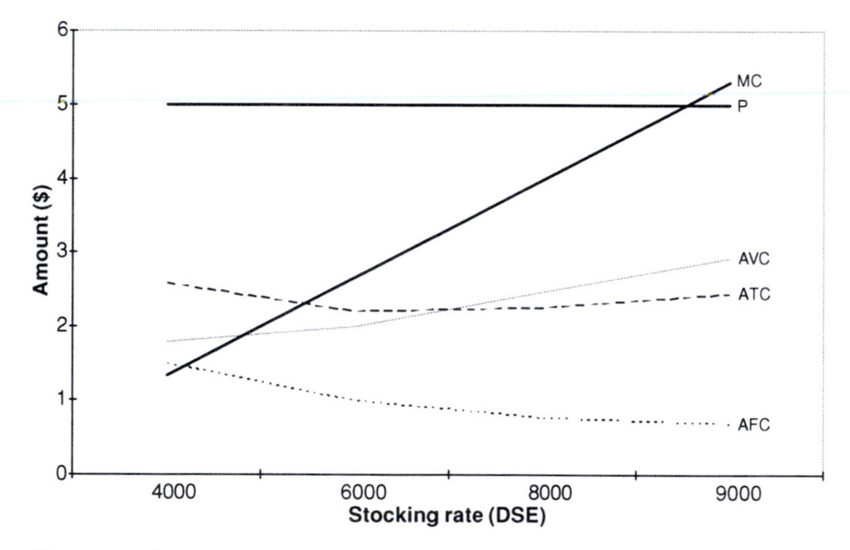


**Figure 3.3** Costs, revenue and maximum profits for a farm

Consider another example.

> Suppose the last application of 20 kilograms of nitrogen per hectare (N/hectare) to a wheat crop cost $10 and the added wheat is expected to be worth $12, but the same 20 kilograms of N/hectare, if added to the pastures, might add 100 kilograms of pasture dry matter worth $20 net return from grazing. If the farmer didn't want to spend more on variable costs because he could not afford it, then he could still increase operating profit by transferring the last 20 kilograms of N/hectare ($20 of variable cost) from the wheat crop (extra return $12) and applying it to the pasture (extra return $20). The farmer gives up the chance of $12 per hectare from wheat and gains $20 per hectare from grazing. The addition to profit is $20 minus $10, leaving $10 extra profit per hectare from adding the last 20 kilograms of nitrogen to pasture, compared with $2 extra profit from using the nitrogen on the wheat. For the same total variable cost, the farmer is $8 per hectare better off.

In the short run, profit can be increased by increasing total gross margin by spending money on variable costs, without affecting fixed (or overhead) costs. The TGM can be increased by spending money on variable costs in various activities according to the criteria of: (a) extra cost versus expected extra return; (b) equal-extra expected returns from applying a particular input to alternative activities; and (c) equal-extra expected return from applying different inputs to particular activities.

One way of tackling the problem of increasing profit is by increasing TGM for the existing level of total overhead cost (see Figure 3.4).

In the above example, to increase profit the farmer spends money on variable inputs such as fertiliser, chemicals, seed, casual labour and drenches, which add more to gross income than the cost of the extra variable inputs. This reduces (dilutes) the fixed costs associated with each unit of output, and thus increases profit. The way increasing output reduces the average fixed cost per unit of output is shown in Figure 3.5.

## Increasing profit in the medium and longer run by changing fixed costs

So far, the focus has been on the situation called the short run, where the main farm resources of land, labour, machinery and livestock are a fixed amount, and we are deciding how much to spend on inputs such as fertiliser, water, chemicals,

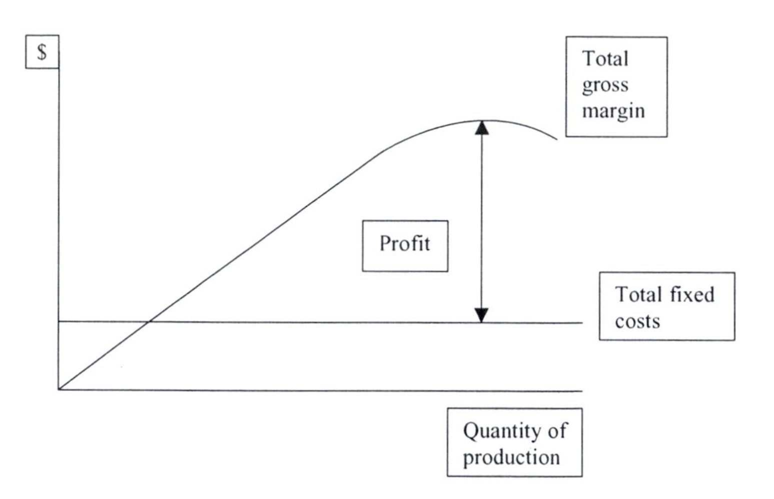


**Figure 3.4** Gross margin, fixed costs and profit

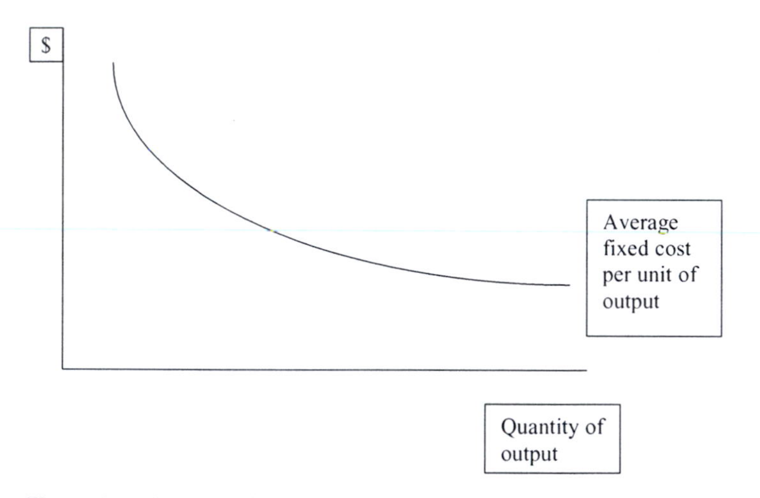


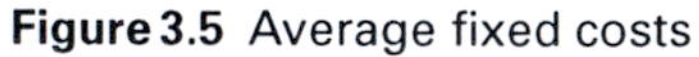

**Figure 3.5** Average fixed costs

seed, fodder (called variable inputs) to maximise profits. As the length of the planning period extends beyond the immediate production period, then the other important inputs that were fixed in the short run – the land, labour, livestock and machinery – can be changed. With the longer planning horizon the focus is on all the costs of production, not just the ones that affect production in the short run. The focus in the medium and longer run is the total size of the whole farm business; in particular, having a size that enables adequate profits to be made. This is the main challenge farmers face – how to expand the business in order to remain profitable as the prices they receive decline in real terms and their costs rise in real terms.

As the length of the planning period extends, there is scope to increase profits by reducing the fixed or overhead costs per unit of output, as compared with the short-run solution of increasing output for a given level of overheads. To a large degree, profit in farming is maintained over the medium and long run by reducing the overhead costs for a given amount of output. The most common means has been spreading the fixed cost of the operator's labour and management over more output.

> Suppose that a cropping operation had $500,000 invested in machinery that was used to crop only 1,000 hectares per year. The machinery had an expected life of 10 years and no salvage value. The annual depreciation would be $500,000/10 years = $50,000. This is $50 per hectare cropped. Now, suppose that the farmland area is expanded and 1,600 hectares of crop was grown each year instead of 1,000 hectares, using the same machinery. The depreciation cost per hectare becomes $31. Some other machinery-related costs would increase, such as repairs and maintenance and casual labour, as the equipment is doing more work. Overall, though, the average total cost per hectare of producing the crop will be less with the larger area than when the smaller area was being cropped.

Over time the increased supply of agricultural products causes the real prices received for agricultural products to steadily decline. In response, some farmers find their overhead costs are too high for their business to be profitable, and they cease farming. This enables other farmers to expand the size of their business by buying or leasing more land, investing in larger equipment, and adopting new technology to improve crop yields and stock-carrying capacity. That is, generally, they attempt to increase the output per business and to spread the fixed costs, and thereby maintain profit despite getting lower real prices for the product they sell. This is the story of modern farming: increase productivity to maintain profit or get used to declining profits.

## Increasing profit through focusing on marginal costs, not average costs

It is common for agricultural consultants and accountants to advise farmers to estimate the average costs of running the farm business. Superficially, this might seem a sensible thing to do. Surely, if the average costs of producing a tonne of wheat is $150 and the price being paid for wheat is $100/tonne, this information tells the farmer something sensible? Well, usually, it doesn't. What happens is that the question gets confounded by the time perspective. An accountant, with hindsight, can estimate what the average cost of producing something was in the past. In doing this, all costs – fixed and variable – are relevant. But economic thinking is about making decisions for *future* production. For the short run, when the farmland, machinery and livestock are owned, and permanent labour is employed, the relevant focus is on costs that can be changed in the coming production period. Another way to think about it is in terms of the costs that can be avoided, and those that cannot be avoided, in the short run. Those costs that cannot be avoided are not relevant to the decision in the short run. Only those that can be changed and which directly affect production are relevant. Thus economic thinking looks only at the extra costs incurred in a planning period and compares these with the extra revenue. Consider the example below.

A horticultural product processing business is producing 10,000 tonnes per year of processed horticultural product which it is selling at $500/t. However, the costs of labour, raw materials, packaging, transport, marketing and so on have increased the average total costs of production to $600/t. The managers know that they cannot sell their product in the market for $600/t, because there are many other processors who can supply the market at $500/t.

Now, a buyer for a large supermarket chain comes along and offers to buy an extra 10,000 t/p.a. at $400/t. That is, the supermarket offered $400/t for products that were costing the business an average of $600/t to produce. Obviously, it would not be profitable for the business to accept this low price offer from the supermarket. Or would it? At present the processor is losing $100/t, which amounts to $1 million per year. Is it true that the only way to become profitable would be to sell the current production at, say, $700/t? What would happen if the processor accepted the offer to buy 10,000 tonnes at $400/t? The answer to this question lies in the marginal costs of production, not the average costs. If the cost of producing an extra tonne is $200 (overhead costs the same as for 10,000 tonnes of output), then the firm could accept the offer of $400/t for 10,000 tonnes. The comparison would be:

Before supermarket offer:

| Output | Price | Total revenue | Total cost | Average cost |
|---|---|---|---|---|
| 10,000 | $500 | $5,000,000 | $6,000,000 | $600 |

If offer taken up:

| Output | Price | Revenue | Cost | Marginal cost |
|---|---|---|---|---|
| Existing 10,000 | $500 | $5,000,000 | $6,000,000 | |
| Extra 10,000 | $400 | $4,000,000 | $2,000,000 | $200 |

That is, if the offer from the supermarket chain to buy an extra 10,000 tonnes at $400/tonne is *not* accepted:

Total sales = 10,000 tonnes
Average cost = $600/t
Total costs = $6m (10,000 tonnes × $600/t)
Total revenue = $5m (10,000 tonnes × $500)
Profit = *TR* – *TC* = $1m loss

If the offer is accepted:

Total sales = 20,000 tonnes
Marginal cost = $200/t
Average cost = $400 ($8m/20,000)
Total cost = $8m ([10,000 × $600] + [10,000 × $200])
Total revenue = $9m ([10,000 × $500] + [10,000 × $400])
Profit = *TR* – *TC* = $1m

How could this happen? The extra cost of producing an extra tonne of processed product was $200, even though the average total cost before the change was $600/t. The extra or marginal cost is lower, however, because the same overhead costs are incurred whether 10,000 tonnes are produced or 20,000 tonnes are produced. The cost of producing one extra tonne is $200, which can be sold for $400/t. Thus a marginal gain of $200/t is made on each extra tonne of output, an overall contribution of $2m to profit. This wipes out the $1m loss made on the first 10,000 tonnes and leaves an overall profit of $1m.

Consider similar logic applied to a farm example.

Suppose that a dairy farm system of 500 cows, each being fed 5.5 tonnes of dry matter/year and each producing 5,000 litres/year, is estimated to have an expected average total cost of producing a litre of milk in the coming season of $0.25. At calving, a milk processor offers to buy an extra 500 litres/cow at $0.20/litre. The cows could produce an extra 500 litres each if they were fed an extra 500 kilograms of hay and grain over the lactation. This would cost $0.15/kg and each kilogram would

> produce an extra litre of milk worth \$0.20. Suppose that the only extra cost involved was the \$0.15/kg of extra feed. If, when thinking about the coming production period, the farmer thought in terms of averages and average total cost of production, then he or she would think: 'My average total cost per unit for 5,000 litres is \$0.25, and the processor is offering \$0.20 per litre for extra milk, so it isn't worth producing the extra milk.' However, this would be wrong. The extra cost of producing an extra litre is only \$0.15, and the extra return is \$0.20/litre. This adds \$0.05/litre for an extra 500 × 500 = 250,000 litres, a gain of \$1,250. When it comes to maximising profits in the immediate production period, marginal thinking gets it right. Average thinking gets it wrong.

The message is clear: extra costs compared to extra returns tells the producer whether to produce more or less output; average costs of a particular level of production don't indicate whether to increase production or reduce production in order to increase profits. When you are already in the business, average cost of production is a flawed notion when it comes to business decisions. The only place for using average cost of production thinking is before the investment in the business is made and when the appropriate size of the business is being considered. At this stage, all costs are avoidable and changeable, so *all* potential costs are relevant. All potential costs divided by all potential output gives potential average cost of production. Comparing the potential average cost of production for a given size of business can be usefully compared with the expected prices per unit of output – before the investment is made and before any costs of any type are incurred.

## ANALYSING BUSINESS PROFIT, LIQUIDITY, GROWTH AND RISK

The essential criteria used in evaluating the health of a business and the merit of a change to a farm business's operations are as follows:

- the nature and amount of total assets controlled, and how they are financed;
- the contribution of activities (activity gross margin) to farm total gross margin and thus to profit;

- actual recent and expected profit and return on the capital invested, after tax and risk is allowed for in the calculations;
- actual recent and expected net cash flows before and after servicing debts;
- the likely break-even criteria when a change is made – that is, the performance required of the 'new' situation for it to be as good as either the existing situation or another option; and
- expected effects on debt and equity of a change as shown in the current and expected end-of-period balance sheet of the business.

Budgets of expected operating profit, net cash flow and a comparison of balance sheets showing change in net worth over time are therefore extremely useful tools for management, and the starting point for understanding the health of a farm business. Budgets are about now and the future. Accounting statements – such as profit and loss, the sources and uses of cash, and the balance sheet or net worth statement – are records about the past. Management is largely about decisions and control, and thus is concerned with the uncertain future. While it is useful to know how the business has performed in the recent past, it is of more use to ponder how it might perform in the future. Of course, the past partly determines the future. The main analytical techniques for understanding the prospective state of a business – the balance sheet at the start and end of a period, and profit and cash budgets – are explained in the sections that follow.

## Balance sheet

Assets are things of value that the manager controls. Debts are financial obligations. The business has to repay funds that have been borrowed. The balance sheet (also known as the statement of financial position) is about the value of assets and debts at a point in time, not over time. The balance sheet records the firm's assets (what it owns and controls) against its debt (funds borrowed from non-owners) or equity (funds put into the firm by owners). Capital is valued at a point in time. The balance sheet summarises the stock of capital involved in the business at a point in time, such as the start of a period for which the business is going to be analysed. The information in the balance sheet reduces to a simple identity: the value of total assets minus the value of total liabilities equals equity (or net worth). Whereas cash flow and profit and loss (also known as the statement of financial performance) statements refer to flows of funds and profits over time, the balance sheet gives a picture of the assets and debts of the business at a particular point in time.

**Table 3.2** Balance sheet

| Assets | | Liabilities | |
|---|---|---|---|
| *Current (short-term)* | | *Current (short-term)* | |
| Cash deposits | 100,000 | Bank overdraft | 200,000 |
| Stock of grain on hand | | | |
| *Intermediate (medium-term)* | | *Intermediate (medium-term)* | |
| Livestock | 500,000 | 4-year loan | 300,000 |
| Machinery | 300,000 | | |
| *Fixed (long-term)* | | *Fixed (long-term)* | |
| Land | 3,000,000 | 15-year loan | 600,000 |
| | | Total liabilities | 900,000 |
| | | Total equity | 3,000,000 |
| Total assets | 3,900,000 | Total liabilities & equity | 3,900,000 |

Assets in the balance sheet can usually be categorised as current, intermediate and fixed. Current assets and current liabilities are short-term assets and liabilities, usually a time period of less than a year. Longer-term assets and liabilities can be considered intermediate (or medium term) or long term (or fixed). Current assets are easily converted to cash – called highly liquid; long-term assets are the least liquid. Liabilities are debts or claims on the assets of the business. The category 'current liabilities' also includes accounts payable. Intermediate or medium-term debt, and long-term debt, is debt for which payment isn't due in the current year. Another category of debt is contingent liabilities. A contingent liability is an obligation that becomes due in specific circumstances. A common contingent liability is capital gains tax, which only becomes due if the capital asset is sold, or a lease payment on an asset that is due to be paid over a number of years.

The market value of the total resources used in a farm business is the total capital value, or the walk-in-walk-out value (WIWO), of the farm. It is calculated by adding the market value of the land, improvements, machinery, equipment, animals, and stocks of product and inputs on hand at some time. The total capital used in a farm business is the market value that would be received if all the land and improvements, stock, machinery and equipment, and stocks of inputs and product that are used were sold. The value of the stock of capital used in a business varies over the period involved. To estimate the amount of capital involved in producing an operating profit over a period of time, such as a year, the average value of all the capital invested in production over the period is used. The average value of capital is estimated as the opening value of total capital plus closing value of the capital, divided by 2. For example:

$$\text{Average value of total capital} = (\text{Opening value of capital} + \text{Closing value of capital})/2$$

Profit doesn't indicate economic efficiency until it is related to the amount of capital used to produce it, expressed as the percentage return on total capital (operating profit/total capital). This indicates the rate of earning of the total capital relative to the rate of earning of that capital if it were employed in some other income-producing activity.

$$\text{Percentage return on total capital} = \frac{\text{Operating profit}}{\text{Average value of total capital (WIWO)}} \times 100$$

Note that when leased land is used in a farm business, as is becoming increasingly common, and the question of interest is how efficiently all the resources are being used, then the value of leased land is included in the total capital being managed. Also, operating profit is estimated before lease payments are deducted. Operating profit is the return on all the resources used and the indicator of efficiency of the whole business. If leased land is included in the balance sheet, the present value of future lease obligations is treated as a liability of the business.

The sum that would be available from the sale, after paying off any debts owed by the business, is the owner's own capital, called equity or net worth. A farmer's equity can be expressed as a percentage of the total resources under his or her control. The equity percentage of the farm described above is:

$$\begin{aligned}\text{Equity percentage} &= \frac{\text{Equity}}{\text{Total assets}} \times \frac{100}{1}\\ &= \frac{2,980,000}{3,420,000} \times \frac{100}{1}\\ &= 87\%\end{aligned}$$

Shareholders' equity is the owners' claims on the firm. Claims on assets come from whoever provided the means used to acquire these assets. However, claims on debt are stronger than claims on equity. Debts have to be met before equity can be returned to the owners.

Solvency and liquidity of the business are the other relevant aspects of the balance sheet. A business is solvent when assets are greater than liabilities.

$$\text{If } \frac{\text{Total assets is greater than 1}}{\text{Total liabilities}} = \text{solvent}$$

A business is insolvent if it is in a state where if it was to be sold, all debts could not be met.

Liquidity refers to the ability to meet all the cash demands that have to be met in the planning period. A test of liquidity is whether cash and near-cash (current bank deposits, government bonds and securities, saleable stocks of grain, wool and trading livestock) will be able to meet the interest on debts and debt repayments when they fall due in the short- to medium-term future.

It is informative to distinguish between changes in net worth and changes in structure of the balance sheet, and the different implications for liquidity. For example, an equal increase and decrease in assets can come from a farm operator buying some asset such as machinery or livestock and paying for it with cash out of the bank. The structure of short-term assets relative to intermediate-term assets has been affected, and net worth is unchanged, but liquidity is changed. Or, taking out a medium- or long-term loan to pay off a number of short-term loans results in an equal increase and decrease in liabilities, leaving net worth unaffected. The debt has been restructured, but the annual debt servicing requirements, and thus liquidity position, has been changed.

## Profit budgets

Profit and loss budgets show the expected returns and costs of the business in the relevant planning period. Operating profit (also called earnings before interest and tax, or EBIT) is defined as:

Gross income (includes cash sales of produce plus non-cash changes in inventories)

*Minus* Variable costs (usually these are cash costs)

*Equals* Total gross margin

*Minus* Overhead costs (includes cash and non-cash costs)

*Equals* Operating profit (EBIT) (the return on total capital)

The returns and costs in the profit budget have cash and non-cash items such as depreciation and inventory changes. The profit budget has all income earned from farm operations and all costs incurred in that period. Cash farm operating expenses include variable and fixed cash costs. Expenditures for the purchase of capital assets are not considered an annual cash expense but an investment, since these assets are used by the business for more than one year. Instead, the cost of capital assets is allocated over their service life by including annual depreciation as the annual cost.

Note that interest payments on loans are not an operating expense. Financing arrangements are distinct from the production activity. Interest payments are to do with financing the business, not production, and are a reward to those who lend capital to the business.

Estimates of profit include estimates of income and costs that are not actual cash receipts or payments. For example, such estimates might include the value of stock produced but not yet sold, or the amount of depreciation of a piece of equipment in a year. Most capital items depreciate in value over time, from wearing out through use or from obsolescence. An allowance for this cost should be deducted from gross income each year so that all the costs of producing output in that year are set against all the revenues produced in that year. The simplest way of calculating depreciation is to use the straight-line method, which has the assumption that an item depreciates by the same amount each year. With this approach the annual depreciation cost is assessed by estimating the current market value of the asset, its expected remaining life, and the expected value of the asset at the end of its life. For example:

Current market value $200,000

Expected remaining life = 5 years

Expected salvage value after 5 years is $50,000 (2005 dollar values)

$$\text{Annual depreciation} = \frac{\text{Market value} - \text{Salvage value}}{\text{No. of years}}$$

$$= \frac{\$200{,}000 - \$50{,}000}{5}$$

Depreciation per year = $150,000/5 = $30,000

In this case, the depreciation cost of the machine in this year is expected to be $30,000 in 2005 dollar values.

The reward for operator's labour and management must be costed and deducted if a realistic estimate of costs, profit and the return on the capital of the business is needed. An indication of the value of an owner-operator's labour and management can be gained from what professional farm managers get paid, or the next best alternative return an owner-operator could expect to earn elsewhere (called opportunity cost).

Operating profit is the return on all the capital used in the business, and is the reward to all who have contributed the capital used in the business. Interest is paid out of operating profit to creditors. Operating profit minus interest is the reward to the farmer's own capital. This is called net farm income or net profit. That is:

Operating profit – Interest paid to creditors (return on lenders' capital)
= Net farm income or net profit (return on owners' capital).

Lease payments are also a financing expense and a reward to the owners of the assets that are being leased. In such cases:

Operating profit
*Minus* Interest
*Minus* Lease costs
*Equals* Net farm income or net profit

Net farm income is available to the owner of the business to pay taxes, or for consumption expenditures above what has already been allowed for in the owner-operator's wages, or for new capital investment, or for some other use such as repaying debt. Often, an owner-operator's wage isn't paid as such, and drawings of cash are simply made from the business as needed. However, to estimate operating profit, a realistic, market-equivalent cost of the labour and management services provided by the owner-operator has to be deducted as a cost of operating the business – that is, the owner-operator's salary. If the owner doesn't actually draw that amount out and consume it, then they are in effect 're-investing' some of their reward for working in and running the business back into the business, in the form of providing resources for repaying debt or financing new investment.

Anything left over from net farm income after consumption (above owner-operator's salary) and tax represents an increase in the owner's total wealth, called equity. What is left over from net farm income after taxes and consumption could be used to reduce a debt, or to have a cash reserve in the bank, or to buy an asset. These all amount to an increase in equity. For example:

Operating profit
*Minus* Interest
*Equals* Net farm income
*Minus* Income tax

*Minus* Personal consumption above operator's allowance (If consumption is less than the owner-operator's salary (allowance) that has already been deducted from operating profit, then add back the operator's allowance and deduct whatever is actually drawn out of the business for consumption.)

*Equals* Change in equity (net worth)

## Farm activity analysis

To estimate business profit, the contribution to profit of each activity needs to be estimated. An activity is a particular type of production – for example, autumn lambing, first cross ewes producing prime lambs, or wheat in a direct drilled, wheat–grain–legume crop sequence, or a Hereford breeding herd producing yearlings, and so on.

The gross margin (GM) of an activity is the gross income generated by that activity minus the variable (direct) costs incurred in earning the income from the activity. Overhead (or fixed) costs don't come into the analysis of the GM of an activity. The GM of an activity is calculated to identify the contribution the activity makes to farm total gross margin. Then, TGM is the sum that is available to pay the overhead costs and make an operating profit. For any set of fixed costs, the sum of GMs from each activity on the farm determines operating profit.

The contribution that individual activities appear to make to farm TGM, and thus to profit, is useful information as long as it is well recognised that *the performance of any activity in a farm plan is affected by the other activities also in the farm plan*. That is, the contribution of one activity to farm profit is in part a function of the other activities also on the farm. Activity GMs calculated on the basis of GM per unit of some resource used – such as land area, or feed supply, or labour used or capital invested – can be compared with a view to expanding or reducing activities. Estimating the contribution an activity makes to TGM can be useful as long as it is remembered that the contribution to TGM of the GM of any single activity would be different with different combinations and sizes of activities on that farm. The whole farm operation needs to be kept in view. Comparison of activity GMs for different farms with different combinations of activities makes no sense. The whole farm approach is used because when the whole farm is broken down into separate activities, some of the 'whole' is lost, and so too is some valuable information that changes understandings and conclusions about the state of the business or the merit of possible changes to the business.

### *Animal activity gross margin*

An animal activity GM is animal income minus variable costs. Income from an animal activity is made up of sales of animal products and profit or loss from trading of animals and changes in animal inventories over the year. Livestock trading profit or loss comes from sales of animals produced during the year, the difference between the value of old animals sold and the costs of replacing them, plus any change in value of the flock or herd due to a change in numbers from unsold production, or to a change in values of animals.

The variable costs of any animal activity are:

- *Feed:* maintenance costs of improved pastures, cost of forage, crops, hay, straw, silage, purchased feed, home-grown grains, agistment, and direct or casual labour costs.
- *Husbandry:* services and other direct operational costs, animal health care costs, animal breeding costs, contract and casual labour services (veterinary, shearing, transport, and so on), identification methods, and repair and maintenance of activity-specific equipment.
- *Marketing:* brokers' and agents' fees, transport, selling charges and levies.

The GM of an animal activity can be calculated as follows.

1 Sales of animal products
2 *Plus* Livestock trading profit or loss from sales of animals produced and an increase or decrease in value of stock on hand from the start to the end of the year. This includes the value of animals not on hand at the start of the year, but on hand at the end of the year because they have been produced during the year. It can also include any change over the year in the value of the animals on hand at the start of the year, as they change from one class of stock (for example, one-year-old heifer) to another class (two-year-old cow). The values chosen here can have a major effect on income – for example, if a current market value was unusually high or low. As investment in livestock is a medium-term commitment, expected medium-term capital values of the animals are used in estimating the profitability of this investment. It is better to use the same values for each class of stock on hand at the start and end of the year. The difference in value between animals that have been culled each year because they are now too old (called cast-for-age animals) or otherwise unproductive or inferior, and purchases of replacement animals each year to

maintain the size and quality of the herd or flock, is annual depreciation or appreciation of animals.

3 (1 + 2) = Gross animal income

4 *Minus* Costs of feed and husbandry services and direct operational costs and costs of marketing.

5 (3 – 4) = GM

To calculate an animal activity GM, it is first necessary to estimate the flows and changes of animals over the year. This can be done with a flock or herd diagram, as shown in Figure 3.6.

## Livestock trading schedule (to calculate trading profit/loss)

The livestock trading schedule contains the changes that occur in an animal activity through the course of a production period, as well as information about flock and herd trading profit or loss for the year. The schedule records all the additions to and subtractions from the initial numbers and value of the animals in an activity for the production period of interest (a production cycle, usually a year). It acts as a check on all of the movements into and out of an activity, as well as a record of changes in values of animals as they are bought, sold, born and age through the year. The livestock trading schedule forms the basis for calculating trading profit from the animals in the activity. Trading profit from animals plus animal product sold makes up total income from the activity. A livestock schedule is shown in Table 3.3. In it is recorded:

- animal type and class;
- number and value at the start of the relevant time period;
- births;
- purchase numbers and value;
- sales numbers and value;
- deaths and rations;
- transfers in and out of classes within an activity;
- transfers in and out of activities; and
- number and value at the end of the time period.

Table 3.3 is an activity budget for a prime lamb activity. A farmer has provided information as follows: a flock of 1,000 Border Leicester–Merino cross ewes

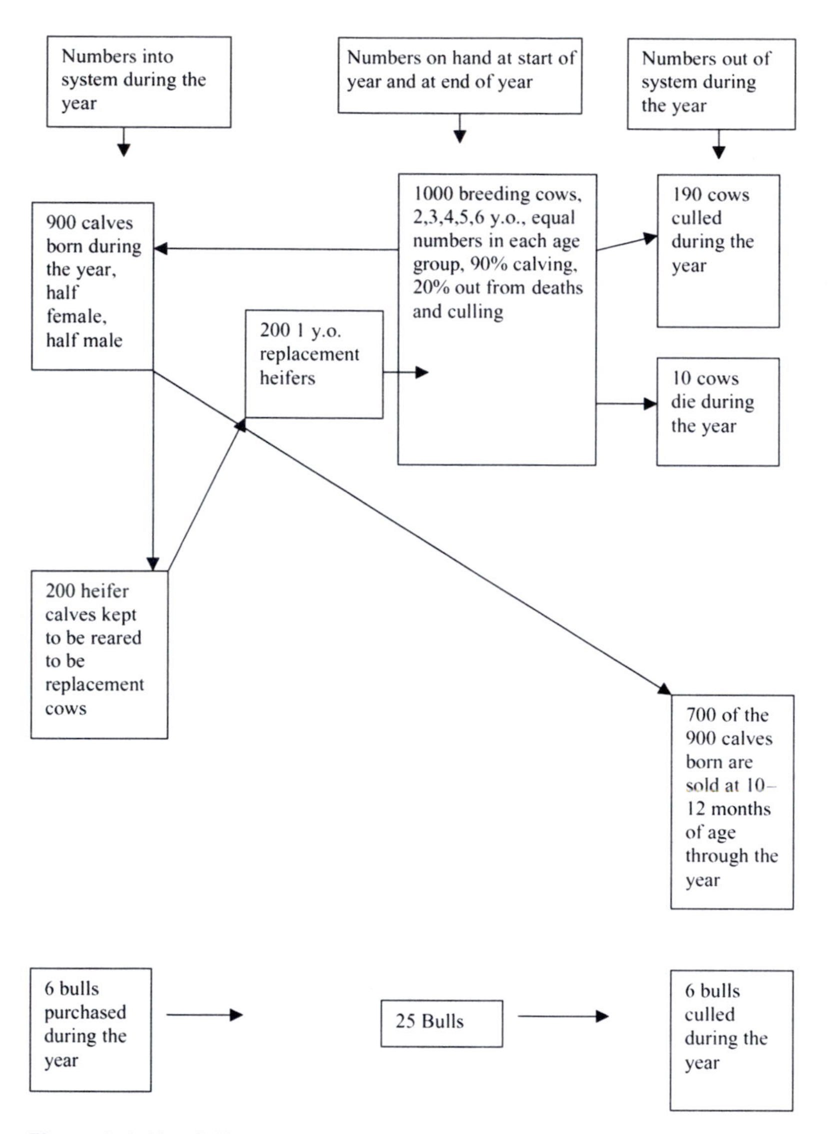

**Figure 3.6** Herd diagram

are mated to prime lamb sires. The ewes are made up of five age groups of 200 head, being two, three, four, five and six years old. Ewes have their first lamb at age two, are kept for four subsequent annual lambings, and are culled, or cast-for-age (CFA), at nearly six years old. The numbers of lambs weaned and sold equate to a rate of 100% of ewe numbers. Ewe deaths are 2% in each age group. Replacement ewes are bought in at nearly two years of age. Replacement ewes

**Table 3.3** Livestock trading schedule: Prime lamb activity

| | No. | $/hd | Total value | | No. | $/hd | Total value |
|---|---|---|---|---|---|---|---|
| *Opening no.* | | | | *Sales* | | | |
| 2 y.o. ewes | 200 | 120 | 24,000 | Cull ewes | 196 | 30 | 5,880 |
| 3 y.o. ewes | 200 | 120 | 24,000 | Lambs | 1,000 | 80 | 80,000 |
| 4 y.o. ewes | 200 | 120 | 24,000 | Cull rams | 10 | 30 | 300 |
| 5 y.o. ewes | 200 | 120 | 24,000 | | | | |
| 6 y.o. ewes | 200 | 120 | 24,000 | | | | |
| Rams | 30 | 400 | 12,000 | | | | |
| *Births* | 1000 | | | *Deaths* | | | |
| | | | | 4/age group | 20 | | |
| *Purchases* | | | | *Closing no.* | | | |
| Rising 2 y.o. | 216 | 120 | 25,920 | 2 y.o. | 200 | 120 | 24,000 |
| Rams | 10 | 800 | 8,000 | 3 y.o. | 200 | 120 | 24,000 |
| | | | | 4 y.o. | 200 | 120 | 24,000 |
| | | | | 5 y.o. | 200 | 120 | 24,000 |
| | | | | 2 y.o. | 200 | 120 | 24,000 |
| | | | | Rams | 30 | 400 | 12,000 |
| Total | 2,256 | | 165,920 (*A*) | | 2,256 | | 218,180 (*B*) |

Trading profit $B - A = \$52,260$

have to replace the four ewes from each age group that die, plus the 196 ewes in the six-year-old age group that reach the end of their productive life at the end of their sixth year, after rearing their fifth lamb. The difference between the purchase price of replacements and the value of the CFA animals is a depreciation cost, or in some animal systems, appreciation income. No account is taken of changes, from start of year to end of year, of the market value of particular classes of stock not sold during the year, because such changes in market values can be transitory and distort the conclusions about income for the year. Changes in market values of a more permanent nature can be captured by periodically revaluing the flock. Changes in the value of animals as they change age classes over the course of the year are counted as inventory income or depreciation costs by the inclusion of the animals in the next age class at the end of the year and valued in closing stock accordingly.

## Prime lamb and beef activity gross margins

Livestock trading profit is estimated and makes up part of activity gross income for estimating activity gross margin (Table 3.4).

**Table 3.4** Activity gross margin: Prime lambs

| Gross income | $ |
| --- | --- |
| Wool | |
| 1,000 ewes @ 5 kg greasy @ $6/kg | 30,000 |
| 30 rams @ 6 kg greasy @ $6/kg | 1,080 |
| Trading profit | 52,260 |
| Total income | 83,340 |
| *Minus* variable costs | |
| Shearing and crutching | 5,000 |
| Animal health | 5,000 |
| Supplementary feed | 10,000 |
| Freight | 3,000 |
| Wool selling expenses | 2,000 |
| Stock selling expenses | 6,000 |
| Annual fertiliser | 8,000 |
| Pasture maintenance | 3,000 |
| Total variable costs | 41,000 |
| Activity gross margin | 42,340 |
| GM/ewe: $42.34 | |
| GM/DSE: $21.67 | |

Tables 3.5 and 3.6 are a livestock trading schedule and an activity gross margin budget for a beef breeding activity producing vealers.

The most useful way to express a livestock activity gross margin depends on the reason for which the calculation is being done. To estimate whole farm TGM and operating profit, total activity gross margin is needed. Livestock activity gross margins can also be expressed as gross margin per hectare, gross margin per head, gross margin per quantity of capital invested, or gross margin per unit of feed used – say, megajoules of metabolisable energy (MJ/ME) or dry sheep equivalent. (A DSE requires 3,000 MJ/ME, or 300 kilograms of 10 MJ/kg dry matter.) The GM/DSE can be used to compare the contribution to TGM of alternative animal activities that utilise the same total feed supply.

## Crop activity gross margins

Expected crop GMs combine expected yield, price and variable costs for crop alternatives. Activity GM is the income from sales of output plus unsold or consumed stocks produced, minus variable costs for a given year. Activity GMs are usually expressed as GM per hectare. The total variable cost of cropping includes:

**Table 3.5** Livestock trading schedule: Beef activity

| | No. | $/hd | Total value | | No. | $/hd | Total value |
|---|---|---|---|---|---|---|---|
| *Opening no.* | | | | *Sales* | | | |
| 2–6 y.o. cows | 500 | 800 | 400,000 | Cull cows | 100 | 600 | 60,000 |
| 1 y.o. heifers | 105 | 500 | 52,500 | Yearlings | 350 | 600 | 210,000 |
| Bulls | 15 | 1,500 | 22,500 | Cull bulls | 4 | 800 | 3,200 |
| *Births* | 455 | | | *Deaths* | 5 | | |
| *Purchases* | | | | *Closing no.* | | | |
| | | | | 2–6 y.o. | | | |
| Bulls | 4 | 3,000 | 12,000 | cows | 500 | 800 | 400,000 |
| | | | | 1 y.o. | 105 | 500 | 52,500 |
| | | | | Bulls | 15 | 800 | 22,500 |
| Total | 1,060 | | 487,000 (*A*) | | 1,060 | | 748,200 (*B*) |

Trading profit $B - A$ = $261,200

**Table 3.6** Activity gross margin: Beef yearling production

| Gross income | $ |
|---|---|
| Trading profit | 261,200 |
| *Minus* variable costs | |
| Ear tags | 3,000 |
| Veterinary and health | 15,000 |
| Supplementary feed | 20,000 |
| Fertiliser | 30,000 |
| Pasture maintenance | 10,000 |
| Livestock selling and transport | 15,000 |
| Total variable costs | 93,000 |
| Activity gross margin | 168,200 |
| GM/breeder: $336 | |
| GM/DSE: $22 | |

- *Growing costs:* such as seeds, fertilisers, water, labour, sprays, and machinery running costs (fuel, oil, repairs).
- *Harvest costs:* such as casual labour, machinery running costs and harvesting materials.
- *Marketing costs:* such as direct costs of storage, processing, transport and selling.

**Table 3.7** Activity gross margin per hectare: Wheat

| | |
|---|---|
| Gross income 3 t/ha @ $200/t | $600 |
| *Minus* variable costs | |
| Freight, levies, etc | $50 |
| Seedbed preparation | $20 |
| Fertiliser: Superphosphate | $50 |
| Urea | $50 |
| Herbicides and application | $90 |
| Sowing | $10 |
| Harvesting | $20 |
| Insurance | $10 |
| Total variable costs | $400 |
| Gross margin per hectare | $200 |

- *Repair and maintenance costs:* to cover costs due to use-depreciation. (Depreciation cost due to time and obsolescence is an overhead cost, and occurs whether the machine is used or not.)

Table 3.7 is an example of an activity GM budget for a wheat activity.

The expected GMs of individual phases of crop sequences are not sufficient information on which to base decisions. The way activities are analysed depends on why they are being analysed. The relative lengths of alternative crop sequences, and the size and timing of GMs of each phase of a sequence, have to be considered to validly evaluate alternatives. The effects of one crop in one year on another crop in another year have to be taken into consideration. These effects might be of benefit, such as providing a disease break or nitrogen. Or, they could be harmful, such as depleting nitrogen, or adding to yield-reducing or cost-increasing populations of disease-causing organisms, weeds or crop re-growth. The GM of a crop activity is specific to the land area under consideration and is affected by the history of that piece of land.

In comparing the profitability of different farm plans involving different crop and livestock combinations, returns from entire sequences are compared, not individual segments of a sequence. If a long fallow has to be used in order to grow reasonable crops, the land has to be set aside for six months or more without producing anything except perhaps some short periods of grazing of unwanted grasses and other weeds. This means that the GM per hectare devoted to crop has virtually to be halved – that is, one fallow hectare and one crop hectare are needed to produce each crop.

If it is assumed that each segment of each sequence will be present on the farm in each year, then the annual TGM per sequence-hectare is the figure to use to compare with alternative rotations. For example, a crop sequence might be one hectare of wheat followed by one hectare of peas. In any year, total crop area is sown half to wheat and half to peas (see below). The profitability of a cropping sequence depends on the GM of the crop phases, the GM of non-crop fallow or pasture phases, and the relative lengths of the various phases of the sequence.

$$\text{Wheat GM/ha} = \$240$$

$$\text{Peas GM/ha} = \$260$$

$$\text{Annual GM/Sequence ha} = \frac{\$240 + \$260}{2}$$

$$= \$250/\text{ha}$$

Generally, when thinking about crop activity gross margins, it is more sensible to think in terms of uses of particular areas of land over the next few years than planning as though the whole farm fits into a routine rotation, with each component present in appropriate proportions across the farm in any single year. The reasons for this are: (a) different land classes on a farm lend themselves to different crop sequences through time; (b) what happens in an area of crop in one year has implications for the activity on that land in the following year(s); and (c) farmers act opportunistically and don't generally rigidly follow fixed crop rotations.

## Using activity budgets in planning a change

If an activity promises a GM of $240 per hectare and an alternative activity has an expected GM of $300 per hectare, it isn't necessarily correct to infer that the most profitable step to take would be to implement the second activity in place of the first activity. Maybe some simple changes could raise the expected GM of the first activity to $300 per hectare, whereas the alternative activity may not have any scope for such improvement. Or, maybe the first activity has complementary effects on another activity that makes it more valuable. It is necessary to first see whether the existing GM of an activity could be improved by the use of better technology or management.

To use GMs for planning the mix of activities on the farm, first select the activity with the highest GM per unit of the assumed most limiting resource (often, but not always, land), and then expand this activity to the maximum level permitted by whatever is limiting more of it being done, such as land, labour, capital, rotation needs, preference for a mixture of activities to spread risks, and so on. Other activities are introduced in order of decreasing GM per hectare until further increases in TGM cannot be achieved without exceeding the limits imposed by the resources available for this activity and any other limiting considerations. Note that activities usually cannot be considered independently. For example, the complementary use of land with crops (such as wheat and sheep) needs to be considered, as does the feed provided by cereal stubble for a sheep activity. If this isn't considered, then the GM of the cereal activity is understated, and the GM of the sheep activity is overstated. GMs are not the correct technique to use if a change in farm plan involves a change in overhead costs, as is most often the case. Then a partial budget is needed (as explained later).

## Costs of machinery

There are a number of types of costs involved in any machinery operation. The first category of costs of owning a machine is called the fixed or overhead costs and includes depreciation, insurance, registration, shedding, and opportunity interest on the capital invested. The capital invested in the machine has an opportunity cost. That is, the money could earn interest in another use. All of these costs are incurred regardless of whether or not the machine is used. The second category of costs is called variable costs. These are direct costs of operating the machinery and include fuel, labour, tyres, lubrication, batteries, repairs and maintenance. These costs are constant per hour of operation and consequently vary directly with the hours of operation. Repairs and maintenance expenses per hour generally increase as the machine ages. The third category of machinery costs is called penalty or timeliness costs. Operations carried out at other than the best times, because of the size or reliability of machinery, can incur a cost to the machine's owner. This cost comes from weather losses at harvest, poorer yields, or a poorer quality of product, than would have been achieved with better timeliness of the key machinery operations.

## Depreciation

Most capital items depreciate in value. This is due either to wearing out or to obsolescence. A depreciation cost has to be counted to get the true cost of the business operations in a year. The simplest way to estimate annual cost of using capital items over a year is the straight-line method. This method has the assumption that an item loses value by the same amount each year. Consider the case where a farmer needs to estimate the depreciation of some machinery in order to be able to estimate his or her annual profit. Suppose that the machine is expected to last five more years and that it would cost $100,000 to replace at the present time. After five years it is expected the old machine will be able to be sold for $40,000 in current dollar terms. The annual depreciation costs are calculated below.

Depreciation cost:
Current replacement cost minus expected salvage value at end of life in current dollars divided by the number of years of expected life:

$$= \frac{\$100{,}000 - \$40{,}000}{5} = \$12{,}000$$

## Interest

Suppose that a farmer was considering investing in a new machine and was worried about having a lot of money 'tied up' in the machine, and thereby incurring a significant opportunity interest cost. The interest cost is estimated for the average value of the capital that will be tied up in the new machine, over the life of the machine. The initial value of the machine, plus salvage value, all divided by 2, gives the average value of the capital tied up over the life of the machine. In this case, if the machine is to cost $100,000 and be worth $20,000 after 10 years, the average value of the capital invested is:

$$\frac{\text{Current value} + \text{Salvage value in current dollars}}{2} = \$60{,}000$$

An interest cost of 7% is charged – that is:

$$\begin{aligned} \text{interest} &= \$60{,}000 \times 0.07 \\ &= \$4{,}200 \end{aligned}$$

Adding the other fixed costs of registration, insurance and shelter gives the total annual capital, interest and other overhead costs of owning a machine. Note that the choice of numbers to use in the calculation of depreciation and interest depends on the reason for doing the calculation, and on whether or not it is

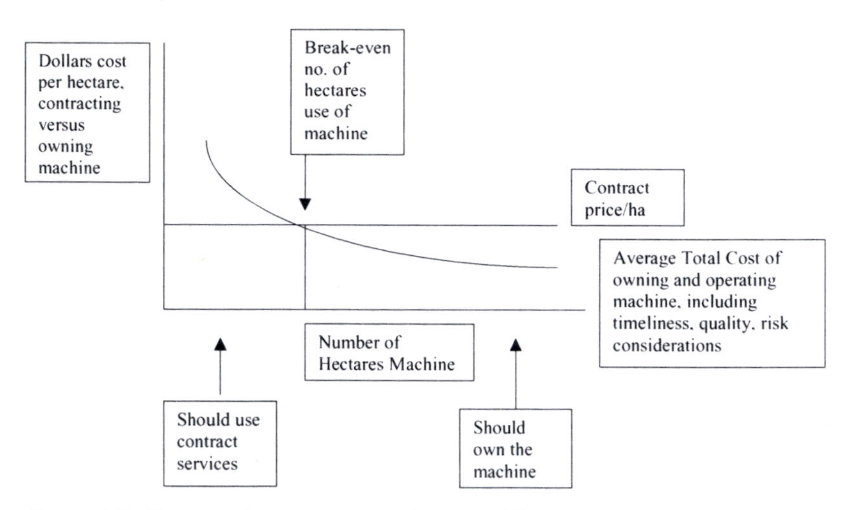


**Figure 3.7** Contracting versus owning machinery

necessary to allow for inflation in the future. To compare costs between alternative machines, all today's dollars and real (no inflation) interest rates can be used. To plan how much to allow so that machines can be replaced at the end of their life, the relevant future inflated dollars and interest rates need to be used. To estimate annual profit of a year's operations, and expected replacement and salvage values, current dollars can be used. For net cash flows, actual interest payments are relevant.

## Annual use

The less a machine is used, the higher the fixed costs per unit (per hectare or per hour). As the machine is used more, the fixed costs are spread over more hours or hectares, so the per unit costs are lower. It won't be economic to own expensive machinery that will have little annual use. Below some level of machinery use it will be cheaper to pay a contractor to carry out the task. Alternatively, above a certain level of machinery use, it will be cheaper to own the machine than to hire contract services (see Figure 3.7).

## Whole farm cash budgets

Net cash flow is the difference between total cash payments and cash receipts in any given period. A sources and uses of cash statement shows cash on hand at the start of the year, where cash came from and went to during the year, and how much was left at the end of the year. This amount then becomes the cash balance for the start of the next year. A cash flow budget has the same information as a sources and uses statement, but being a budget, it is about the future – where money might come from and go to in the coming year.

Expected annual cash surplus estimated in a whole farm cash budget is useful for management as an estimate of the cash that might be available to pay all the expected bills and also to put to various 'new' uses. Cash from all sources is counted.

For example:

Expected annual gross cash receipts from sales of production (excludes non-cash income) plus other cash from such things as off-farm work or new borrowings.

*Plus* Cash variable and cash overhead costs (excludes non-cash costs). (This might include a cash allowance set aside to replace equipment.)

*Minus* Other cash uses (such as interest, principal repayments, tax and consumption spending).

*Equals* Expected annual cash surplus (available for 'new' uses).

The amount of annual cash surplus remaining after interest and loan replacements, personal expenses, necessary replacement of assets, taxes, and new investment from current income shows how much new debt could be serviced.

Probably the most useful financial management tool is the monthly cash flow budget, budgeted out for a reasonable planning period, for example, 12–24 months. Then, as time rolls on, the actual expenditures and incomes are checked against the expenditures and incomes that were expected, on which plans were based. Comparing actual versus expected is a good control measure, enabling early warning of significant changes in performance and facilitating appropriate responses.

## Growth

Owners of businesses are usually as concerned with growth in wealth (measured as change in net worth or equity in the balance sheet) as they are with annual profit, return on capital and net cash flow. Growth is constrained by external factors such as tax and external credit rationing by financiers, and by internal factors such as internal credit rationing resulting from 'used-up' borrowing capacity or withdrawals to finance personal consumption.

The formula for estimating growth is:

Operating profit

*Minus* Interest

*Equals* Net farm income

*Minus* Taxation

*Minus* Consumption or drawings that exceeds any owner-operator's allowance previously deducted in estimating operating profit

*Equals* Growth in equity

This growth shows up as the difference between net worth in the balance sheet at the start of the period and net worth in the balance sheet at the end of the period.

Growth in equity in the farm business is directly affected by:

- rate of return on total resources (operating profit);
- interest on total debt;
- debt to equity ratio (called gearing or leverage);
- rate of personal consumption of net farm income; and
- rate of tax.

Credit can affect the growth of a farmer's wealth over time if it is used to increase the earnings of the farmer's resources. The increase in earnings must exceed the full costs of borrowing, and allow the principal on the loan and interest to be repaid. The relation between debt and equity, and the rate of return on borrowings, govern what the farmer will eventually be worth.

## Putting it all together: The links between profit, cash and growth

Changes in operating profit and in annual cash surplus from one year to the next end up as changes in the assets, liabilities and equity components of the balance sheet. A simple illustrative example of the links between balance sheets, and profit and cash budgets follows.

A farm business is in the position shown in Tables 3.8 to 3.11.

**Table 3.8** Balance sheet at start

| Assets | Liabilities |
|---|---|
| $5m | $2m 10-year term loan just started<br>10% p.a. interest<br>$5m – $2m = $3m Equity<br>Equity = $3m/$5m (60%) |
| $5m | $5m |

**Table 3.9** Profit budget

| | |
|---|---|
| Gross income | $1.5m |
| – Variable costs | $0.5m |
| = Total gross margin | $1.0m |
| – Overheads (including depreciation $100,000) | $0.5m |
| = Operating profit | $0.5m |

**Table 3.10** Cash budget

| | |
|---|---|
| Cash in | $1.5m |
| – Variable costs | $0.5m |
| – Cash overheads | $0.4m |
| – Principal | $0.2m ($2m × 1/10) |
| – Interest | $0.2m ($2m × 10%) |
| New capital investment | $0.1m |
| Net cash flow | $0.1m |

*Growth*

Operating profit $500,000

– Interest $200,000

= Net farm income $300,000

– Tax 0

– Consumption 0

= Growth $300,000

**Table 3.11** Balance sheet at end

| Assets | Liabilities |
|---|---|
| $5m – $0.1m depreciation + $0.1m cash from net cash flow surplus + $0.1m value of new capital investment = $5.1m | $2m – $0.2m principal repaid = $1.8m<br>$5.1m – $1.8m = $3.3m equity<br>Equity% = $3.3m/$5.2m = 65% |

What has happened?

- Assets have increased by $200,000.
- Debt has reduced by $200,000.
- Equity has increased by $300,000, from $3m to $3.3m, a growth of 10%.
- Equity percentage has increased from 60% to 65%.

## Gearing and growth and the principle of increasing risk

Financial risk is a mixture of:

- interactions between operating profit and interest on debt;
- effects of profits and losses on asset and equity values; and
- effects of gearing and interest rates (debt/equity) on debt servicing ability and liquidity.

When considering net profit, return on own capital, liquidity, growth, business survival and bankruptcy, the balance sheet feature of most interest is the ratio of debt to equity.

$$\frac{\text{Debt}}{\text{Equity}} = \text{Gearing}$$

With a high debt to equity ratio, the risk of going bankrupt is also high. When operating profit doesn't exceed interest costs, equity has to be used to pay the interest. The proportion of assets ($A$) to debts ($D$) is commonly expressed as equity ($E$) percentage – that is:

$$E = \frac{A - D}{A} \times 100$$

A high equity percentage means a relatively low proportion of debt (low gearing), and hence a greater margin of total assets over total liabilities. This means less vulnerability to insolvency due to a fall in asset values. In Australian farming, equity percentage for the majority of farms is usually 85% or higher. At different stages of the life of a business, equity will be down to 70% or even 50%. In most farm businesses it is difficult to sustain low levels of equity, because at the relatively high levels of debt there will be many times when the cash available to service the debt will be insufficient.

Consider the following situation.

> A farm business has total capital of $10m, made up of $5m debt and $5m equity. Operating profit is 10% p.a. on total capital, an operating profit of $1m. Interest costs 8% p.a. on $5m, so $400,000 interest is paid. Net profit is $600,000. If there is no tax or consumption uses of net profit, net worth or equity increases by $1m – $400,000 = $600,000; a growth rate of $600,000/$5m = 12%. Now consider the alternative situation where the business makes a 10% p.a. operating loss, a loss of $1m. The business still has to meet its interest obligation on the debt of $400,000. The total loss is $1.4m. This represents a decline in equity of $1.4m/$5m, or 30%. The rate of growth of equity when things went well (+12%) isn't symmetrical to the rate of decline of equity when things went badly (–28%); hence the principle of increasing risk.

The principle of increasing risk is about the obvious phenomenon whereby as the debt of a business increases relative to equity, the more risk there is of that business not being able to meet its debt servicing obligations at all times. It is also about the phenomenon that if businesses have a certain amount of debt and use the debt capital such that the return on total capital exceeds the cost of the debt, then the owner's equity grows at a faster rate than would have been the case if the business had a lesser amount of debt. The principle of increasing risk explains why businesses cannot simply keep borrowing and growing. The reason is that as the proportion of debt in total capital increases, so too does the risk. So, as debt increases, capacity for further borrowing decreases; at the same time, the cost of further borrowings increases because it is more at risk. The business faces a trade-off between borrowing and equity growing rapidly when things go well (earnings exceed cost of debt) and putting equity at risk of declining when earnings are less than the cost of debt, and with equity declining at an even greater rate than it grows when things went well. Hence, increasing debt means increasing risk. The further complication is that if a business doesn't use debt, and use it well, it won't grow sufficiently rapidly to remain competitive and equity will eventually be lost as the business is forced out of business.

## Loan repayments

Loans require repayments of principal (the sum borrowed) as well as interest payments. The two main types of loan repayments apply to term loans and amortised loans. With a term loan, the principal is repaid in equal instalments (annual, half-yearly, quarterly, monthly). Interest is paid separately on the outstanding principal and thus reduces as the principal is repaid. Term loans are a common form of financing provided by the trading banks. Term loans are usually either for farm development or farm purchase, and are generally available for from five to 15 years. Amortised loans are loans where the repayments are made in equal instalments consisting of both principal and interest, called an annuity. As the principal is repaid, the interest content of each payment is reduced, allowing more principal to be repaid. To calculate the annuity (equal annual payment) that repays an amortisation loan, use the formula for an annuity whose present value is 1 – namely:

$$\text{Annuity} = \frac{P[i(1+i)^n]}{(1+i)^n - 1}$$

where $P$ = the principal sum borrowed

$i$ = interest rate

$n$ = number of years

Alternatively, use discount tables (see Appendix) where the formulas for calculating annuities (such as the formula above), and for present and future values of sums of money, have been calculated and are expressed as discount 'factors'. For a $100,000 loan over 10 years at an interest rate of 10% (from Table D, Appendix), an annuity whose present value is 1, the discount factor is 0.1627. This discount factor comes from:

$$\frac{0.1(1+0.1)^{15}}{(1+0.1)^{15} - 1} = 0.1627$$

Thus the annuity that will repay the principal and interest on a $100,000 loan is $100,000 × 0.1627 = $16,270 annual repayment. This annual repayment is made up of both interest and principal. In year one of the loan the interest component is $100,000 × 10% = $10,000 and the rest, $6,270, is principal repayment. In year two the principal owing is $100,000 – $6,270 = $93,730. So, interest in year two is $93,730 × 10% = $9,373 and the principal repayment is $16,270 – $9,373 = $6,897, and so on. Loan repayments should be matched to expected repayment capacity, which usually increases as a farm is developed. When a business is in financial difficulty, one of the first things to look at is the possibility of restructuring existing debt commitments to better fit expected improvements in annual cash surplus.

## Taxation

Taxable income of a business has to be assessed each year. In analysis for farm management decision making we are concerned with taxation matters as they affect the numbers to put in farm budgets and the impact on the outcomes of farm management decision making. In this regard, there are always tax considerations to take into account. However, this is quite different from the annual analysis which accountants carry out to minimise the tax a business has to pay. The information provided each year for tax purposes is of little value for informing farm management analysis.

Taxable income is gross income minus deductions of expenses, but it is different from operating profit because different measures of some income and expense items are used. Income tax in Australia is a 'progressive' tax, so the more one earns the greater the rate of tax paid. A set rate is charged for a range of different levels of income, called income brackets. The higher the income bracket, the higher the tax rate. Companies are taxed at a flat rate of 32 cents in the dollar. Partnerships don't pay tax. The individual partners are taxed on their share of net income.

Allowable deductions for taxation purposes are operating expenses, depreciation and some capital expenditures, such as for land and water conservation. Depreciation of machinery and equipment is worked out on the basis of effective life as estimated by the Commissioner of Taxation. At times there are concessionary allowances as well, such as an extra 10–20% of this value as an accelerated rate of depreciation for the life of plant and equipment. Structural improvements are depreciated for tax purposes at about 3–4% per year.

Capital expenditures that are deductible vary, but these often include investment in land and pasture improvements that can be written off at 10% p.a. over 10 years. At times, special concessions such as a 100% write-off for tax in the year of expenditure are available for things such as investments in water conservation and land preservation. Other concessions to farmers include rebate of diesel fuel, and sales tax exemptions on specified classes of vehicles and on many inputs used in agriculture, such as fertilisers, and so on.

Tax legislation in Australia also takes consideration of the variability of farming incomes and the risky Australian climate. It has a number of helpful provisions for farmers, such as averaging of taxable income, accelerated depreciation allowances, income equalisation deposits, the treatment of fodder for stockfeed, and the spreading of proceeds of forced sales of stock over a period. Farmers' incomes can be subject to tax averaging. 'Averaging of taxable income' means that the farmer's current income is taxed at the average rate of tax that applies to the farmer's average taxable income over the previous five years.

Ultimately, most changes in the operation of a farm affect the tax payable. An estimate of the tax implications of changes is usually needed in assessing the merit of a change to the business. It isn't good enough to compare expected profit before tax with alternative actions, because different actions have different implications for tax. For instance, for a proposed investment, some of the capital expenditures might be tax deductions in the first year Another investment might have few tax-deductible expenditures and might be subject to capital gains tax.

In estimating net benefits from a change, the extra taxable income is approximated as gross income (including retained animal natural increase at concessional values) minus deductible extra operating expenses, allowable extra capital deductions and extra interest. Extra taxable income multiplied by the approximate marginal tax rate of 15–20% gives the extra tax payable.

Tax accounting is misleading for the purposes of farm management for a number of reasons. The main ones are:

- Different costs are used in calculating taxable income and operating profit. For example, depreciation cost is derived from the Taxation Commissioner's standard 'effective life' of a machine plus a concessionary extra percentage, and is based on historic cost. For management purposes, operating profit uses depreciation based on the expected life of the machine and the replacement cost.
- Different values of animal inventory income are used. For taxation accounting, for 95% of businesses, livestock are valued on the basis of annual cost and concessional values for natural increase. Estimates of operating profit use market values of livestock.
- For tax purposes, some capital expenditures are depreciated at accelerated rates.
- Livestock values for tax purposes don't equal the actual amount of capital invested in livestock.

The important things to remember with taxation are: (a) there is always a tax angle; (b) employ a good tax accountant; and (c) emphasise actions to increase net income, rather than tax-decreasing ploys. The marginal tax rate is never 100%.

### *Returns after tax*

The real return after tax is what counts with any action. An investor receiving a nominal 15% p.a. return when inflation is 10% and the 40% marginal tax bracket applies gets a true return after tax of 15% minus (15% × 0.4) = 9%. Take away the effects of 10% inflation and the real return is –1%. If paying 25 cents in the dollar tax, and inflation is 8%, then to receive a real return after tax of 4%, a nominal return of 4% real plus 8% for inflation, giving a 12% return after tax, is needed. To receive 12% nominal return after 25% tax, an investor needs a pre-tax return of 16% nominal.

### *Livestock and tax*

There are a number of special provisions under the tax Act (*Income Tax Assessment Act 1936*) to deal with livestock income. To assess and to pay income tax it is necessary to identify what income has been earned, and it follows that it is necessary to know the value of livestock on hand at the end of the year. Two methods can be used to value the animals on hand at the end of the year: the average cost or market value. Using the market value method requires that numbers of animals in various classes be known and their current market value be assessed. The average cost method of valuing livestock at the end of the year is based on the opening value plus purchases plus natural increase. A concession given by the Commissioner of Taxation is that the annual natural increase of livestock can go into the trading account at concessional values chosen by the operator. The concessional method of valuing natural increase causes the average value of the livestock for tax purposes to be pulled down over time. Average cost of the livestock will tend towards the value selected for natural increase. But when animals are sold, their market value making up sales income is used for tax purposes. The difference in the total amount of tax paid using either market values or average cost of stock on hand comes through the timing of the tax payments being different, which can be a significant factor.

A major management implication of income tax on livestock income is the way the choice of method of valuation can affect the animal replacement system used. The calculation of average value of livestock at the end of the year means that purchases at market prices are added to the opening value of the stock. This adds more to the average value of the head/flock than does natural increase. Natural increase goes into the calculation of average value at less than market rates. Raising the average value per head by purchases means that the value of closing stock numbers is higher, and thus trading profit is higher, than when average value and closing stock value is lower (owing to low values for natural increase). Thus the tax-deferring effect of concessional values for natural increase (by keeping average value of head down) is greater than for purchased livestock. The tax system works in favour of having self-replacing flocks and herds over using a bought-in system of replacing animals.

### *Capital expenditures*

The tax deductibility of the annual costs of capital investments of the farm effectively reduces the real cost of such investment. For example, a farmer invests $50,000 on pasture development. This represents an annual tax-deductible expense of $5,000 p.a. for 10 years. If the average marginal tax rate (that is, the marginal rate applying to the typical five-year average) was 35%, then the $5,000 deduction would amount to $1,750 savings in tax ($5,000 × 0.35). When we talk of $1,750 benefit per year, this is the saving in tax payable from using resources in the way identified. The tax paid, and saved, associated with an alternative use of the $50,000, is also relevant in weighing up the decision to invest in one way or another.

### *Machinery*

For tax purposes, machinery depreciation is based on effective life of the machine. A 20% tax allowance for machinery depreciation has the following effects. Say the machine costs $10,000 to buy. The 20% tax deduction for depreciation from taxable income reduces taxable income by $2,000. The tax saving isn't $2,000 but the saving in the tax that would have been paid on this amount had it remained as part of taxable income, instead of being deducted from it. The tax saving is the deduction from taxable income ($2,000) multiplied by the average tax rate, say 25 cents in the taxable dollar. This comes to a tax saving of $500 for each year until the machine is written off for tax purposes.

### *Interest*

Taxation affects the way in which resources are allocated to alternative uses. Treatment of costs such as interest as a tax-deductible expense makes capital investment relatively more attractive than would be the case without tax deductibility of interest. For an investor on a high tax rate, say 50 cents in the dollar with interest as a tax-deductible cost, the true cost of borrowing at 10% interest rate is 5% after tax. The tax deductibility of interest makes for a higher rate of growth of equity for a profitable business than would be the case without tax deductibility.

### *Growth*

Tax also has implications for growth of equity. Tax reduces the rate at which equity grows. Tax reduces the growth rate in a profitable year, particularly with a progressive tax system. But it doesn't affect the size of a loss in a loss year. Therefore, tax accentuates the effect of the principle of increasing risk, which has the effect that a loss reduces a highly geared farmer's equity more quickly than it would grow after a profitable year.

### *Capital gains tax*

Since 1985, a tax has been levied on inflation-corrected capital gains. The tax is charged at half the individual's marginal tax rate. Farm transfers within the family are exempt from capital gains tax, as are farm sales that are re-invested in a farm.

### *Taxation and leasing*

Usually lease payments for machinery and livestock are tax deductible, subject to some minimum residual values set by the Commissioner of Taxation. If the lease arrangement has the condition that the lessee will buy the leased asset at the end of the lease, the lease payments are not allowed as a deduction. Instead, for tax purposes, the transaction is regarded as a sale, and deductions for depreciation and for interest are allowed. These are similar to lease payments.

### *Tax and budgets*

The question of what tax rate to use in a budget is always difficult to answer. Suppose farmers paid income tax on each year's income, regardless of previous years' incomes. The relevant effect of any change in farm plan would be the addition to annual taxable income from the change. In this case the marginal tax rate would be the right rate to use in farm management budgets. However, averaging provisions are available to all primary producers, and most average their taxable income. For taxation purposes the relevant effect of a change in farm plans, costs and income is the effect it has on the five-year moving average taxable income in any year. The question is then how much a change in farm plan will shift the five-year average taxable income. The effects of extra income will not be felt fully until it has been part of taxable income for every year of the averaging period.

There are many other matters affecting whether or not a farmer pays much tax. The deductible costs start in the first year and may extend for up to the next 10 years. For the above reasons, it is hard to pinpoint all the effects of a change in farm net income on the farmer's average taxable income. The correct tax rate to use is the 'expected average marginal tax rate'. The true 'average marginal tax' rate to apply to the effect of the change is hard to define. So, a guess is needed about the average rate of income tax a farmer might expect to be paying over the next few years. In most cases, farm marginal tax rates are below 20%.

### *Income tax minimisation*

These are the main ways to reduce the income tax paid:

- splitting of incomes when additional taxpayers (usually family members) can be allowed to share in the taxable income;
- forming partnerships;
- setting up trusts – discretionary units;
- forming a company;
- paying a salary to family members for services rendered; and
- maximising allowable deductions by means of:
  - investment allowances;
  - self-managed superannuation funds;
  - tax-exempt deposits such as farm management bonds designed to assist farmers to cope with the variability of their incomes; and
  - income tax schemes which create 'artificial' losses.

Minimising the tax bill is a specialist field. Farmers pay accountants to advise them on how to keep their tax bills down.

## Business organisation

The common categories of business organisation are sole proprietorship, partnership, and company or corporation. Sole proprietorship is a business owned by a single individual. The owner keeps title to the assets and is personally responsible without limit for liabilities incurred. The proprietor is entitled to the profits of the business but has to bear any losses. Partnerships have more than one owner, and all partners are fully liable for debts incurred by the partnership. The partnership agreement sets out the rules by which it will operate, such as the nature and amount of capital to be invested by each partner, how the profits and losses are to be shared, how to add a new partner, or how to reform the partnership in the case where one partner wants to get out. A company is a legal entity that has a life separate from the owners. Ownership is in the form of shares and these are transferable. The owner's liability is limited to the amount of investment in the company – that is, limited liability. This means that creditors cannot call on the shareholders' personal assets to settle the company's debts. The owners elect a board of directors who, in turn, choose people to operate the company.

The distinguishing feature of a sole proprietorship and a partnership is that of unlimited liability. No distinction is made between business assets and personal assets. The distinguishing feature of a company is that it continues, whoever are the particular owners. Companies must comply with company law. Companies have advantages when it comes to recruiting new capital, such as limited liability, ease of transferring ownership through sale of shares, and flexibility in dividing the shares.

The main determinants of the appropriate legal form of business are often taxation considerations and issues to do with intergenerational succession and family inheritance. For sole proprietorships and partnerships, the owner reports the business profits as personal income and is subject to personal income tax. Companies are taxed as separate and distinct identities. The tax rates of individuals relative to those of companies are an important consideration in determining the appropriate form of business organisation.

### *Analysing a business*

It is common for advisers to businesses and consultants who don't bring economics as the core discipline to their analyses to advocate and apply an approach to analysing businesses known as comparative analysis, or benchmarking. The essence of this approach is the belief that it is possible to learn much about what to do in one business by having a close look at the average performance of other businesses engaged in similar activities. While it is true that comparing and contrasting one business with another can provide useful insights into different technical processes used to achieve particular ends, there are many limitations to comparative analysis and benchmarking when it comes to identifying problems and the true cause-and-effect relationships, and deciding on appropriate levels of measures of performance for any particular business. These limitations arise because all businesses comprise a unique mix of resources. First, the people operating the businesses have different goals, skills, attitudes, experiences and family situation, and stage of life. A second source of variations relates to the quantity and quality of resources available to different farm businesses. Third, farming is different from most other businesses in the exposure to volatility and lack of control over key determinants of outcomes. The most useful comparison and contrast for farm businesses is with their own performance over time. There are far more useful approaches to analysing a business, and identifying its strengths, weaknesses and the key contributors to performance measures such as profit, liquidity, growth and risk. Understanding the critical human, technical, economic, financial, risk and institutional factors that determine the performance of a modern business and the environment in which it must operate is more useful than focusing on a few partial performance indicators.

Non-economic business analysts like to compare ratios of technical measures of average performance and to use various crude indicators, such as gross income or units produced, as 'proxy' indicators of profitability, net cash flow and growth. Neither of these approaches is valid for farm businesses.

Superficially, it seems useful to look at one farm that produces, say, an average of 100 kilograms of wool per hectare, or an average of 7,000 litres of milk per dairy cow, or carries on average two cows to the hectare, and then to compare these figures with what happens on another farm. However, this approach is theoretically flawed. Maximising average technical ratios of output to input isn't the means to maximise profit. Indeed, relying on this sort of information can lead to drawing logically opposite conclusions about what to do. For example, maximising the ratio of milk per cow would involve reducing the number of cows per hectare; maximising the ratio of milk per hectare would involve increasing the number of cows per hectare and reducing the amount of milk produced per head – logically opposite conclusions. The correct analysis of how to increase or maximise profit is to have a stocking rate of cows per hectare and a level of production per cow at which the marginal cost of producing an extra unit of output – by whatever means – just equals the marginal return from an extra unit of output. The focus of the analysis and the decision is on expected extra return and extra cost from a change to the whole input–output make-up of the system – not an average of one technical component. Technical ratios tell nothing about the economic state of affairs (see Malcolm 1990; Ferris and Malcolm 1999).

If benchmarking of technical productivity ratios between different systems involving different people with different resources, goals, attitudes to risk, stage of life, and so on, isn't a meaningful exercise, what about comparing measures of economic and financial performance between different businesses? The figures about production, costs, income, capital and debt of a business lend themselves to the formation of ratios that summarise aspects of the business. These 'summaries' are used to compare with other similar 'summaries' about businesses. The most valuable use of comparisons of business summaries or ratios is to compare selected ratios of performance of a business over a recent time period, and to assess expected performance over the forthcoming near future production periods. That is, the business is compared with itself at different times. As outlined previously, the least useful approach is to compare a business with other, different businesses. This is because the detailed situation and resources of any business are unique.

Note also that for ratio analysis to be meaningful, the measures need to be meaningful. Crude aggregate measures, such as using gross sales or output as some proxy indicator for profit or net cash flow or growth, are not as informative as summary ratios that account for both the income and the costs. Agriculture is so volatile that gross measures can be associated with good or bad net results!

The gearing (or leverage) ratio measures the relationship between debt and equity of the farm business. The higher this ratio, the larger are the outside claims on the business relative to the equity of the business. Lenders have a particular interest in this gearing ratio, as it is one measure of the financial risk involved. Another ratio of interest is the ratio of fixed cash cost to total cash costs. The higher the value of the fixed expense to total expense ratio, the less flexibility the farm operator has to adjust quickly and efficiently to a changing market condition.

The main limitation of financial ratios is that they are historical averages, whereas economic theory is about marginal concepts and the future. Economic theory tells that if firms wish to maximise profits, they should operate at the point where marginal costs are expected to equal marginal revenue. As ratios are based on averages they can easily be misinterpreted. Businesses could have the same rate of return on capital because they have the same average costs in producing output and they sell at the same price. But one operator could have a low rate of return compared to the potential return because the size of the business is too small and not producing enough, and another operator could have a low rate of return because the business is too large and is operating inefficiently. A ratio sometimes used in evaluating efficiency of livestock production is average return above average feed costs. Maximising this ratio suggests that farmers should produce at the point where average feed costs per unit of output are minimised. This is the point where average variable costs are at a minimum. However, this isn't the profit-maximising point of production. Total profits could be increased by producing more units of output until the extra revenue from extra output just equals the extra cost.

While there is no limit to the summary ratio measures that can be, and usually are, produced about a business, only a few are really useful in telling the owner much of value about the business. The main ones are measures of efficiency (for example, operating profit as a percentage of total capital), liquidity (for example, net cash flow before and after debt servicing, interest rate coverage ratio), growth (for example, change in equity, gearing ratio) and financial risk (gearing ratio). Trends in such measures over the recent past for a particular business are informative. More useful is the expected movements in the important measures over future planning periods. Comparison of actual and expected business performance, especially cash flows, is also very useful information.

To sum up: The focus on historical average performance isn't useful to management that is concerned with future changes and the extra benefits and extra costs of the change. It isn't possible to deduce the marginal consequences of a change to a farm system from historical average measures of performance – let alone deduce these for one business by having a good hard look at another business. Further, a manager achieving what looks like poor results could be doing a terrific job with terrible resources; another manager in another situation might look to be achieving good results and might be actually doing a terrible job with terrific resources. Simplistic comparisons of average benchmark measures of performance between different farm systems – league tables of top, middle, bottom, best, worst, and so on – tell little about what is actually being achieved and nothing about why it is so. There is an implied cause and effect between a measured average phenomenon and an observed set of actions, which may or may not be the true explanation of cause and effect. The further implication is that if some other business does similar things to the benchmark system, similar results will be achieved. The perennial farmer response to benchmarking data is a querulous: 'And so what? Now what do we do?' The only way to determine true cause-and-effect relationships and to solve business problems is to know the business and the people well, walk the farm, understand the technology and the technical possibilities, and have mastery of the economics, finance and risk analyses. That is, the only way to provide sensible answers to the question 'And now what do we do?' is to properly analyse the expected benefits and costs of feasible potential changes to the system. This is a process that is way beyond benchmarking. Above all, the essence of successful farming is implementing new technology in changed systems. There are no benchmarks for innovations.

In the next chapter, analysis of innovation in the whole farm business is explained.

## CHAPTER SUMMARY

*Economic understanding of an agricultural business has to be built on a sound foundation of technical knowledge. This knowledge is converted to economic information using budgets, and budgets are about the future. Further, marginal and future, not average and past, ways of thinking about the operation of a business are crucial to sound analysis of business decisions. Profit, cash and growth are different phenomena that, taken together, indicate the health of a business.*

## REVIEW QUESTIONS

1. Identify, for a farm system, some examples of applications of the principle of diminishing marginal returns of a variable input applied to fixed inputs.
2. Identify, for a farm system, some examples of applications of the principle of equi-marginal returns when applying several variable inputs to fixed inputs.
3. The economic way of thinking comes down to three main ideas: diminishing marginal returns, equi-marginal returns and opportunity cost. Explain these concepts.
4. Explain why profit is maximised when the marginal return from an extra unit of a variable input just exceeds the marginal cost of an extra unit of that input.
5. Why is it fundamentally important when analysing a business to distinguish between fixed costs and variable costs?
6. Over a defined time, a business can be profitable but have inadequate net cash flow, or it can have good net cash flow but not be profitable. How can this be so? How do profit and net cash flow differ?
7. Average cost of production is a flawed concept that leads to wrong conclusions about how to operate a business. Explain why marginal, not average, thinking is needed for sound business decisions.
8. The difference between the capital components and the annual operating components of a production system is a fundamental distinction to make in order to estimate profit truly. Explain this statement.
9. The health of a business is determined by (a) economic efficiency (profit, return on capital), (b) financial viability (net cash flow after debt

servicing), and (c) growth (change in net worth). What are the income and costs 'equations' for profit, growth and net cash flow? Construct a balance sheet at start of year, an expected profit budget, an expected net cash flow budget, an expected growth budget and expected balance sheet at end of year, for the coming years of a business that you know well.

**10** There are many ways to construct budgets. The shape and form of a budget doesn't matter much. What matters is the logic underlying the relationships between the numbers that go into it. Good judgment about the numbers helps, too. Explain.

**11** Sometimes once a budget is constructed the decision maker thinks the job is done. However, this is when the work *starts*. Budgets have to be 'worked over' – different numbers tried out and possible situations explored – in order to provide information to help inform the decision. Give an example of how a budget might be used to provide new information to a decision maker.

## FURTHER READING

Boehlje, M.D. and Eidman, V.R. 1983, *Farm Management*, John Wiley & Sons, New York.

Campbell, K.O. 1944, 'Production cost studies as a field of research in agricultural economics', *Journal of the Australian Institute of Agricultural Science*, vol. 10, pp. 31–7.

Ferris, A. and Malcolm, B. 1999, 'Sense and non-sense in dairy farm management economics', *Australasian Agribusiness Perspectives*, vol. 2, paper no. 31, www.agrifood.info.

Heady, E.O. 1952, *Economics of Agricultural Production and Resource Use*, Prentice Hall, New York.

Hopkin, J.A., Barry, P.J. and Baker, C.B. 1999, *Financial Management in Agriculture*, The Interstate Printers and Publishers Inc., Danville, Illinois.

Malcolm, L.R. 1990, 'Fifty years of farm management in Australia: Survey and review', *Review of Marketing and Agricultural Economics*, vol. 58, pp. 24–55.

Mauldon, R.G. and Schapper, H. 1970, 'Random numbers for farmers', *Journal of the Australian Institute of Agricultural Science*, vol. 36, pp. 279–84.

# 4
# Analysing innovation in the whole farm business

The main challenge faced by managers of farm businesses is to manage change. In this chapter the main budget techniques for analysing the economic merit of investing to change a farm system are explained.

## INTRODUCTION

Business managers can either embrace change to increase productivity and achieve the necessary growth of their business, or have other less desirable change forced upon them. Having established the state of a business as it currently operates, the main task for management is to analyse the options for change to increase the productivity of the business. This process involves (a) identifying innovations, (b) imagining alternative futures, and (c) judging alternative futures against criteria of feasibility, likelihood and contribution to achieving goals.

Innovation in farm businesses means identifying and implementing different ways of using resources in farm businesses. Analysing decisions about alternative ways of using resources in a business involves using information to help form judgments about relationships between costs and benefits in the changed system sometime in the future, even though much about the future situation is unable to be known well or at all, or is unable to be quantified well or at all. Nevertheless, it is still useful to approach the question of innovation in the business in a rigorous and systematic manner, making explicit what is known and likely and assumed, and thinking hard about, and defining, plausible possible future states of the world with and without the changes in question.

The first difficulty encountered in analysing and forming judgments about questions of future resource use is to establish the appropriate perspective. The correct perspective is that of comparing and choosing between alternative futures within the 'sphere of effects' of the investment. The person conducting the analysis must identify and quantify cause-and-effect relationships linking alternative actions with the identifiably likely possible outcomes.

The analyst has to envisage alternative future states of the parts of the world that will be affected by alternative uses of the resources in question. The decision maker's choice is between Alternative Future Number 1 and Alternative Future Number 2, and so on – not between the status quo and the future. In a dynamic world, by definition, the current situation cannot be an option for the future. As the world changes around the business, the business changes with or without the help of management.

Most of the difficulties that create problems in investment analysis relate in some way to ensuring that the comparisons of alternative futures are valid comparisons. That is, the main challenge is in defining plausible scenarios for the state of the part of the world that is the focus of the investment decision for the alternative situations, both without the change and with the change. From the infinite possibilities, a small number of logical, plausible alternative states of the world need to be imagined and defined. The decision is then based on the expected differences deemed likely to exist between these states in the way criteria are met to do with feasibility, likelihoods and contribution to achieving goals.

It is impossible to know for certain what will happen in the future, and the approximation of the future becomes less and less useful the further out in time we go. So how, then, do decision makers make decisions about the future? First, as they cannot avoid doing so, they must imagine the future. Decision makers unavoidably have to imagine plausible future situations, and analyse how they and their business might look in such situations. If the look is good, and there is a strong likelihood that the actual circumstances of the future will bear a reasonable resemblance to the future they have imagined, then decision makers have a basis for deciding and proceeding and implementing the change in question.

Adapting the business to changes is the core of the management task. Changes to farm plans can be relatively uncomplicated and immediate – for example, a change in an aspect of an activity such as plant variety or cultural operations, or mix of animal activities. Or changes can be complicated, involving significant capital investment that takes time to reach full fruition, and having implications for many aspects of the farm's operations. With relatively simple changes, the changed farm plan is nearly fully operational in a relatively short time, such as within a year or two years, and other aspects of the business are not greatly affected by the change. Where a simple change is contemplated, there is no need to go through as many detailed steps in analysing it as is the case for more complex changes. A budget for one year of operations in the 'steady state' will usually suffice. This is the common partial budget (explained later). Large and complicated change usually takes time before the new activity is fully operational, and has a number of significant implications for the way the whole of the farm operates. Most often, a significant change involves relatively large investment, takes several years to be fully operational, and results in changes in the way that other parts of the farm operate. Examples include a decision to improve an area of low-producing pasture, to install some irrigation, to purchase and develop more land, or to change cropping sequences or the timing of lambing or calving.

A useful distinction to make is for the term 'economic analysis' to refer to analysis of the efficiency (profitability) of resource use after counting all benefits and costs, regardless of the form of these benefits and costs. 'Financial analysis' refers to the liquidity or financial feasibility of a project. The focus of financial analysis is on annual and cumulative net cash flows, and on interest and principal repayment arrangements. Seeing investment in these two distinct ways – economic analysis and financial analysis – helps to avoid one of the most common problems in investment analysis, which is where the economic (total benefits and costs, cash and non-cash) and financial (strictly cash) aspects of the proposal are confounded, often achieving the ignominious outcome of getting wrong the conclusions about both the economic and the financial prospects of the investment.

## ANALYSING INNOVATION

The following sections explain the main questions that need to be answered in order to decide whether a change to a business is likely to be a good move or not.

## Start with the market

The first question to answer when considering any changes to the farm operation is: 'What are the likely market prospects for the change in output that will result from the changed operation of the business in the imagined future?' The market rules. Farm product has to be sold at a price that will cover costs and leave sufficient profit to justify the investment. The decision maker needs to investigate when and how to sell the product(s) resulting from the change, and the likely supply of and demand for the product. The obvious first step in the process of changing resource use in a farm business is to weigh-up the market prospects for the particular output involved, in terms of quantity, quality and timing, and the prices that might be received. The next step is to form a judgment about the quantity, quality and timing, and the costs of producing the changed output in the business. The human and technical feasibility aspects are explored in depth, and the net economic benefits and financial and risk aspects of the change are estimated and pondered.

## Human aspects

The goals of the farm owners are the key to decisions about adopting innovations in their businesses. The goals of the farm owners help to establish the rationale for making the change. The way the change will contribute to the owners achieving their goals needs to be convincingly identified and understood.

The farmer and the employees need to have the interests, skills and knowledge, and the right incentives, for a change to succeed. Also, the appropriate supply of labour must be available when it is needed. The stages of life of the members of the farm family – starting out, children being educated, expanding, winding down, and so on – are critically important in determining what actions are appropriate for any particular business.

## Technical knowledge

A sound knowledge of the technical aspects of a proposed change is necessary. The physical aspects to consider are whether the resources are available (labour, soil type, water, animals, pasture, machinery) to carry out the change, or whether extra resources will have to be obtained. Technical needs may include specific types of fertiliser, specific animal husbandry techniques, and the methods of growing, harvesting and marketing. A detailed physical plan of the land, machinery, crops and types of animals is required. The technical basis of farm budgets has to be correct. Otherwise, all other analysis is a waste of time. The farmer's detailed knowledge and experience of production on the farm is one of the main sources of the detail that is needed. Experimental results from research investigations have to be adjusted to be relevant to the specific situation of each farm. Yields used in budgets have to be those that are possible from using economic levels of inputs on the particular farm.

The important technical considerations when planning a change involving cropping and pasture are:

- How will soil fertility be maintained?
- What sequence of crops and pastures will be followed?
- What methods of soil preparation, weed control and sowing are to be used?
- How will machinery services be acquired – ownership, leasing, contract, share?
- What is the soil nutrient status, and what types and quantities of fertiliser will be required?
- If machinery is owned, how and when will it be replaced?
- What is the typical pattern of pasture dry matter production over a year?
- What are the expected annual maintenance fertility and weed control requirements of pastures?
- If developing an area of pasture, what is the timing of the rate of increase in herbage production, the expected steady-state production, and the corresponding increase in animal carrying capacity?
- What is the expected life of the pasture before major renovation will be needed?
- What are the target levels of key indicators of soil fertility such as soil acidity/alkalinity (soil pH) and available phosphorus?
- What amounts of lime will be needed to achieve the desired soil pH? Will lime be added annually or as a large quantity every 10 years or so?
- How much phosphorus will be needed (a) to raise soil phosphorus to the optimum, and (b) annually to maintain soil Olsen P at this level, given the soil type and dry matter production and stocking rate?
- How will crop residues be utilised?

Some commonly used technical guidelines about fertiliser and pastures are given below.

- A megalitre of irrigation water can be used to produce 1 tonne of utilised dry matter (1.4 tonnes with superior grazing management).
- On some soil types, 10 kilograms of phosphorus might increase soil available Olsen P by one unit; while on another soil type 7 kilograms of phosphorus might be required for a one-unit increase. Depending on terrain, a kilogram of phosphorus will be needed to replace the phosphorus removed by one DSE of grazing in a year – that is, a maintenance P rule is: 1 kg P/DSE stocking rate. However, this can be slightly less on terrain that has less 'run-off' effect.
- An irrigated dairy pasture can provide 7–15 tonnes of dry matter utilised by the cows, depending on the pasture and the stocking rate and the skill of the grazing management.
- A tonne of lime on soil pH can decay at a rate of 5–10% p.a., depending on soil type.

Important technical considerations when planning a change involving animal activities are:

- How are herd or flock numbers maintained?
- What will be the timing of the breeding and production cycle? (This is determined largely by the seasonal pattern of feed supply.)
- What is the pattern and quantity of feed demand?
- What is the pattern and quantity of feed supply?
- What will be the relationship between quantity of feed supply and quantity of animal product (wool/hectare, milk/head, liveweight turnoff/system, and so on)?
- What are the animal husbandry needs (health, breeding)?
- How will animal energy demands vary through the year to meet the requirements of maintenance, breeding, pregnancy, lactation, growth, and product qualities?
- How will energy demands be met throughout the animal production cycle, in the face of a seasonal pattern of energy supply?
- How will energy supply be matched to energy demand, and how will energy deficits be met?
- What are the animal nutrition requirements apart from energy, and how will they be met?

- What is the planned pattern of build-up in herd flock numbers?
- How is the required annual genetic improvement of animals to be achieved?

Note: It makes no difference, in an economic sense, if extra animals required as part of a pasture development program come from retained animals (less surplus animals sold) or from bought-in animals. It is sometimes thought that retained animals are somehow free! They are not, as they could have been sold; thus the cost is the opportunity cost of retaining them and not selling them.

### *Measures of feed demand and supply for different livestock activities and feed supplies*

There has been considerable confusion and ambiguity about the use of standard measures of feed demand of animal activities, and feed supplied by alternative feed sources, such as DSE, livestock months, cow days, and megajoules of metabolisable energy (MJ/ME). The requirements of animals for maintenance, growth, pregnancy, lactation and exercise have been reasonably defined by scientific experiments, and the ME context of sources of feed similarly defined. As explained in Chapter 2, a DSE is the maintenance requirements of a 45 kilogram wether for a year. The assumed rating of DSEs applied to different animal activities can have significant implications for the conclusions drawn. The most precise standard unit used to estimate feed demand and supply is MJ/ME. Information about typical patterns of seasonal pasture production for regions is becoming increasingly available.

As information on MJ/ME required by animals and supplied by feed sources is readily available, it is probably best to work analytically in these terms, while recognising that there remain unknowns about aspects of the assumed relationships between energy required and animal output. Similarly, energy supplied by a feed source is affected by the level of production and utilisation of dry matter that is achieved. This depends on grazing management and animal combinations – for example, sheep and cattle have different patterns of grazing. Grazing activities are dynamic. The adage is: 'Pasture production determines stocking rate, and stocking rate determines pasture production.' Further, note that estimates of feeding standards are not without ambiguity. Much remains to be learned about energy supplied to animals from mixtures of feeds and the so-called associative effects of energy use by animals. Some typical, unsophisticated, rules of thumb about feed demand and supply are set out below.

- A kilogram of dry matter (DM) of 10 MJ/ME energy above animal maintenance requirements might enable a medium-sized dairy cow to produce 1 litre of milk, on average, through a lactation.
- A medium-sized dairy cow producing 5,000 litres of milk per year will need about 5.5 tonnes of DM for maintenance, exercise, pregnancy and lactation.
- A 45 kilogram wether will require about 8 MJ/ME/day for maintenance, or about 300–350 kg/DM/year with 8–10% MJ/ME/kg.
- A breeding ewe rearing a lamb will require as much feed as two DSEs.
- A breeding beef cow rearing a calf weaned at eight months will require as much feed as 12–15 DSEs.
- A 300 kilogram steer grazing pasture that provides 10 MJ/ME/kg/DM will require 10 kg/DM/day (8kg/day for a 200 kilogram steer, 12 kg/day for a 400 kilogram steer).
- A 500 kilogram British breed cow in moderate to good condition will require 10 kg/day of pasture of 10 MJ/ME/kg/DM when dry or pregnant. In early lactation, she requires 15 kg/DM/day, and in the fifth month of lactation, 20 kg/DM/day of pasture that provides 10 MJ/ME/kg/DM. (This includes an allowance for the calf.)
- A tonne of dry matter of 10 MJ/ME will feed about 30 DSEs over a year.
- In spring in southern Australia, around 100 days of average pasture growth at a rate of 40 kg/DM/day is common, under typical grazing pressures with good pasture and well-managed grazing rotations. That is, 100 days × 40 kg/day gives 4,000 kilograms of DM produced in spring. Note that production is affected by stocking rate, and stocking rate is affected by production.
- A hectare of non-irrigated pasture can produce 10 tonnes of DM/ha, of which 5 tonnes may be utilised and supply enough feed for about 15 DSEs, depending on the pasture utilisation rate achieved, which in turn depends on the grazing system.
- In a grazing beef activity, about 7 kilograms of average DM will be converted to a kilogram of bodyweight.
- Annual feed demand and supply are crude measures. A better approach is to work in megajoules of energy, and to break down the energy requirements of animals, or the energy supplied by pastures, into daily or monthly sub-periods within a year.

## Benefits and costs

The key criterion in judging whether an innovation is worth adopting concerns the extent to which the innovation is likely to help the owners achieve their goals. Three significant goals are: (a) making profitable use of resources, (b) having sufficient cash to meet needs, and (c) increasing net worth over time, within a risk context that is acceptable. The contribution an innovation is likely to make to achieving these goals is an important criterion by which to judge the potential change. The measure of profitability is the expected return on the capital invested. The expected (risk-weighted) return on the extra capital invested is the first criterion by which to judge the potential change. Expected ability to finance the change (service the debt) is the second criterion. The third criterion is the expected effect on net worth over the life of the change or the planning period.

### *Analysing a simple innovation: A budget of a single production period, steady state*

A simple change to a farm can be analysed by constructing a budget of the operation of the farm as it currently operates, and then comparing how it would operate with the adoption of an innovation, once the change is fully operational. This situation is called the steady state. If the business looks likely to be more profitable with the change than without the change, then the change is likely to be worth making. As well, the net cash flow after the change has to be adequate to cope with the added financial demands resulting from the change. The addition to net worth likely to result from the change has to look attractive.

There are two ways of investigating the situation of a farm business with change and without change. One way is to construct annual budgets of the whole farm for the situation without the change and for the situation with the change, and to compare expected operating profits for the two situations. Another, more common technique for evaluating a simple change is to look at only those parts of the whole business that will change. This partial approach involves looking at all the favourable aspects of the change and balancing them against all the unfavourable aspects that result from the change. As many as possible of these elements are given a dollar value. Extra costs and extra returns and net gain are estimated to indicate how much whole farm profit is expected to increase after the change. Regardless of the nature of the change, the impacts have to be assessed in terms of the effects on the whole farm. The marginal way of thinking dictates that the perspective to use when considering a change is: 'What might be the situation if the change isn't made?' compared to 'What might be the situation if the change is made?' Take the case of the simplest change: a new activity is added without any change to the existing farm business. The merit of a simple change such as this to a farm business can be assessed using the partial budget approach, and by looking at the situation of a typical production cycle, such as a year of operation of the change performing at the expected 'steady state' level, and:

- estimating the value of all the expected extra costs that can be given a monetary value;
- estimating the value of all the extra benefits that can be given a monetary value;
- subtracting the expected extra costs from the expected extra benefits to get net benefits;
- expressing the expected net benefits as a percentage of the extra capital invested to give the expected return on extra capital;
- comparing the expected return on extra capital with the potential returns from investing the same capital in some other use; and
- investigating the effect of changes to key variables such as yields and prices on the key criterion.

In a partial budget, the extra profit from a simple change is expressed as a percentage return on the additional capital invested. This is done for the time when the effects of the change are fully realised and the plan is fully operational. This takes no account of the costs incurred between initial set-up and steady state, or the effect of time on cash flows up until the steady state. (If the change is complex and involves more than a year or two to reach steady state, the steady-state partial budget is inappropriate. Budgeting over the years of the life of the change and net cash flow analysis and discounting are needed.)

In a more likely case of a simple change than the case discussed above, innovations are likely to affect some of the things currently being done on the farm. For example, in the case where a new activity replaces an existing activity, there are likely to be some net benefits from the 'old' activity that will no longer be generated. There will be some benefits forgone, some costs saved, and some net benefits given up. These effects must also be counted in evaluating a change to a farm system. The approach in this case is:

- estimate the expected net benefits from the new activity (*A*);
- estimate the net benefits from the 'old' activity (*B*) that will be forgone if the change is made;
- deduct (*B*) from (*A*) to get (*C*), the net gain from the change;
- express (*C*) as a percentage of the extra capital invested in replacing the old activity with the new activity; and
- check the effects of variability of key variables on the results.

It can be a little tricky to analyse a change using a partial budget approach when one activity is replacing another. Partial budgeting is used to tell if there is a net gain or loss from change. It doesn't identify the return on all the capital involved in the activity; it only shows the gain from the extra capital invested – that is, the return on the capital required to add the new activity, minus the capital made available because it is no longer required for conducting the activity that is being replaced. If return on total farm capital is of interest, then a whole farm budget, not a partial budget, is necessary. If capital is released from an activity that is being replaced by a change, then it may be that there is some scope to invest it and earn interest on it, and this is an extra benefit of the change.

The variable nature of the weather and of agricultural markets means that no farm plan is likely to work out as expected. So, it is necessary to work out what would happen if prices or yields or interest rates were either worse or better than expected, and the breakeven levels of key parameters. The percentage chance (or probability) of these events happening can be evaluated.

## A partial budget

A farmer wishes to analyse the idea of developing 200 hectares for growing irrigated lucerne. The following details are provided:

- Total farm area is 2,000 hectares.
- The property is all improved pastures carrying 12 DSE/ha, a total of 24,000 DSE.
- A first-cross prime lamb flock of 8,000 head (16,000 DSE) and a self-replacing Merino flock of 4,000 (6,000 DSE) is carried. There is also a small herd of 100 Angus cows producing yearlings (2,000 DSE).

- Expected medium-term GM/DSE from a typical DSE of all the livestock activities is $20 GM/DSE. (This includes the annual pasture fertiliser cost.)
- The capital cost per DSE of a livestock breeding unit is $50/DSE.
- An area of 200 hectares close to a river is suitable for growing irrigated lucerne.
- Expected costs (most likely, current dollars):
  - Land forming for irrigation costs $1,000 per hectare. This is expected to need re-doing at a cost of $300 per hectare at the end of six years. That is, the initial capital investment of $1,000/ha will have depreciated to be worth $700 by the end of year 6. This represents a total depreciation cost of $300 over the six years, or an annual cost of $50/ha.
  - Lucerne seed sown in year 1 lasts six years. Re-sowing is then needed, at 12 kg/ha @ $6/kg – that is, an annual cost of $1/kg.
  - Pumping operating costs/year: $15/ML × 12 ML/ha.
  - Tractor and other plant operating costs/year: 14 hrs/ha @ $20/hr.
  - Water used/year: 12 ML/ha @ $15/ML.
  - Fertiliser costs/year: $100/ha.
  - Spray costs/year: $50/ha.
  - Twine costs/year: $6/t @ 13 t/ha.
  - Extra labour cost/year: $10,000.
  - Tractor and plant depreciation/year: $10,000.
  - Other plant costs/year: $2,000.
  - Fencing costs: $5,000 lasting 10 years; an annual cost of $500.
- Expected income (most likely, current dollars): 11 t/ha/year from year 2 (9 tonnes first year) @ $200/t on farm, baled, in paddock. Lucerne stand will produce 11 t/ha for six years, then taper off. Expected income is 11 t/ha × $200 × 200 ha = $440,000.

Capital invested in the irrigation activity will be as follows. (Good secondhand machinery will be purchased.)

| | |
|---|---:|
| Tractor | $100,000 |
| Scarifier | $15,000 |
| Combine | $25,000 |
| Baler | $35,000 |
| Pump | $40,000 |

| | |
|---|---:|
| Fencing | $5,000 |
| Land forming (200 ha × $1,000/ha) | $200,000 |
| Lucerne seed | $14,400 |
| Total capital | $434,400 |

The marginal tax rate is 20 cents in the dollar. The farmer has the capital needed to make the change and it is currently earning 5% p.a. (real) before tax.

The task is to construct a partial budget and analyse this proposal, based on one year of production at a steady-state level. To do this, calculate real return on marginal capital after tax.

Extra income from irrigation in a steady-state year:

11 t/ha × $200/t × 200 ha = $440,000

Therefore, total extra benefit (*A*) = $440,000.

Extra costs from irrigation in a steady-state year:

- Land forming cost annualised: ($1,000/ha – $700/ha)/6 = $50/ha × $200 = $10,000
- Lucerne seed cost annualised: $14,400/6 = $2,400
- Pumping: $15/ML × 12 ML/ha × 200 ha = $36,000
- Tractor and plant operating: 4 hrs/ha @ $20/hr × 200 ha = $16,000
- Water: 12 ML/ha × $15/ML × 200 ha = $36,000
- Sprays: $50/ha × 200 ha = $10,000
- Fertiliser: $100/ha × 200 ha = $20,000
- Extra labour: 3,000 hrs × $20/hr = $60,000
- Twine: $6/t × 11 t/ha × 200 ha = $13,200
- Tractor and plant depreciation: $160,000/8 yrs = $20,000
- Fencing: $5,000/10 = $500

Therefore, total extra cost (*B*) = $224,100.

Net benefit (*A* – *B*) from production of irrigated lucerne is:

$440,000 – $224,100 = $215,900.

The net benefit forgone from the grazing activity being given up in order to grow irrigated lucerne is:

200 ha × 12 DSE/ha × $20 GM/DSM = $48,000

The net benefit from the change is the net benefit from the new activity – irrigated lucerne – minus the net benefit from the activity that is being given up – grazing. This is:

$215,900 – $48,000 = $167,900

This is before extra tax. With a marginal tax rate of 20 cents in the dollar, extra tax will be $33,580. Net benefit after tax will be $134,320.

The critical question to ask in judging this potential investment is: 'What is the extra return on the extra capital invested?' To estimate this indicator, the capital aspects of the proposal need to be considered. The capital released from grazing will be:

12 DSE/ha × $50 DSM × 200 ha = $120,000 of livestock capital

The added capital required for irrigated lucerne will be:

$434,400 – $120,000 = $314,400

The extra return on capital, after tax, is $134,000/$314,000 = 42% real.

What does this mean? It means that the extra capital invested in the farm business is expected to earn around 40% p.a. in the scenario where 11 t/ha of lucerne hay is produced and it is worth $200/t.

If lucerne hay production is only 10 t/ha and is worth an average of only $160/t, the return on extra capital after tax is 15% p.a. real. If production was only 9 t/ha, and price/tonne was $150/t, there would be no profit in it. Given the risk, a 15% p.a. real return after tax is considered the minimum return required for this investment. Thus, 10 t/ha and $160/t are the minimum acceptable or breakeven levels of these key variables.

The most likely situation is that 11 t/ha will be produced and price will be $200/t. At this, the expected real return of 40% p.a. after tax looks very attractive compared to other opportunities to invest on the farm, or elsewhere in the economy, and is markedly better than the safe real opportunity return of 5% p.a. The breakeven yields and prices seem relatively safely achievable compared to what is most likely to happen. Even if all the funds required were to be borrowed at, say, 7–10% real cost, the project is attractive. Note that the interest payments would save some tax payment by increasing tax-deductible expenses and reducing taxable income. In this analysis, this effect is captured to some extent by using a marginal tax rate of 20%, a rate which could apply after all extra tax deductions are included in the estimate of taxable income.

Note that a partial budget is used to look at one year of operations typical of the way the investment is likely to perform over the life of the investment. Partial budgets are most useful for evaluating changes on the farm that take place over a relatively short time and so there isn't much time lag between the initial outlay and the returns coming in. Once significant time is involved, then the effects of time on the value of money have to be counted.

## Time effects and discounted cash flow budgeting

The value of a defined quantity of benefits or costs, measured in dollar terms, and received at different times in the future, is different at these different times in the future. This is true even if there is no inflation in the economy that changes the purchasing power of a dollar. The reality is: people prefer a dollar today to a dollar sometime in the future, because the dollar received today is worth more to them than a dollar received in the future. The value of a dollar of benefits or a dollar of costs at year 1 and year 2, or any other future time, is different, even without any effects of inflation. A dollar of benefits or costs today is worth more than the same sum at any time in the future.

Today's dollar invested can grow to be worth more than a dollar in the future. A dollar today is equal to $1.10 in a year if it is invested and earns 10% for the year. Or a sum less than a dollar today is equivalent to a dollar sometime in the future, because the less-than-a-dollar today can be invested to grow to a dollar sometime in the future. That is, $0.90 today is the equal of $1 in one year if it could grow at a rate of 11% interest over the year. So, a dollar worth of benefit, or cost, received in year 1 and year 2 or year 10 isn't the same in value. The sooner a dollar is received, the sooner it can be put to work to earn and to grow in value. This means, in practice, that when counting the benefits and costs from an investment, the dollars of benefits and costs in year 1 are not considered to be the same in value by the investor as the dollars of benefits and costs at, say, years 2 or 5 or 10.

In order to be able to add all the benefits and costs of an investment when the benefits and costs occur over a number of years, the dollars received and spent in each of the years (dollars that are not of equivalent value to the investor because they are received at different times) have to be adjusted to the same value terms. These future, different-valued dollars are converted to their equivalent value at some common point in time. This is usually the present time. Future dollars are converted to their equivalent present dollar values. (Alternatively, present dollars can be converted to the same value as a dollar received in the last year of the investment, called future values.) Once the benefits and costs over the life of the investment are all expressed in equivalent dollar values, then they can be added together meaningfully. (Before adjustment, the future dollars received or spent at different times were like apples and oranges; unable to be sensibly added.) Once the streams of future benefits and costs are expressed in the same value terms and can be added, the total costs can be subtracted from the total benefits, to give net benefits. This sum, net benefits, is called the net present value (NPV) of the investment.

Alternative projects will have different patterns of cash flow over their life. Such projects need to be compared to see which project will contribute the most to the net worth of a business. Once the dollar values of benefits and costs in the different projects are adjusted to present values, and the net benefits of the alternative projects are estimated, then the net benefits of each of the alternative projects can be summed and compared. The net benefits of the projects are expressed in the same value terms. The project with the highest net benefit is the best.

Adjusting the value of future benefits and costs to equivalent present values involves using a technique called discounting to present value. This value-adjusting technique is a way of allowing for the fact that benefits and costs occurring at different times have different values. It also allows for the fact that when resources are invested in one use, the chance of using those resources in another use is gone. Discounting means deducting from a project's expected benefits and costs the amount the invested funds could earn in the most profitable alternative use. The rate of adjustment of future dollar values to present dollar values is called the discount rate. The discounted value of future dollars is calculated as follows:

$$PV = FV/((1 + r)^n)$$

where $PV$ = present value of the future amount of a benefit or cost expressed in dollars

$FV$ = future amount of a benefit or cost expressed in dollars

$r$ = the opportunity cost rate of earning in another use, expressed as a decimal

$n$ = the number of years it will take to receive the future amount of a benefit or cost expressed in dollars

The formula $((1/(1 + r)^n)$ gives a discount factor that can be multiplied by the future value to convert it to a present value. For example, at 10% opportunity cost rate of earning in another use, the discount factor to use to adjust a future dollar to its present value is $((1/(1.1)^1) = 0.909$. Multiply $\$1 \times 0.909 = \$0.91$. A dollar in one year is adjusted to \$0.91 equivalent present value. There are discount tables where the calculations of discount factors $((1/(1 + r)^n)$ are already calculated for particular combinations of years and interest rates (see Appendix, Table 2).

To take the opposite approach and find the future value to which current dollars will grow, the discounting formula is rearranged to get:

$$FV = PV/((1 + r)^n)$$

This is the formula for compound growth of money. A dollar today grows by the compound growth factor of $((1 + r)^n)$ – that is, the compound growth factor of 1.1, to equal \$1.10 in a year's time. Compound growth factors are shown in Appendix, Table 1.

As a first look at an investment proposal, the common method of appraisal is the expected rate of return on capital in a typical year. This is an approximation only of the true return to capital over the entire life of the investment. This might not matter as long as the method ranked correctly the choices under consideration. However, the only theoretically correct, and most accurate, method of determining what an investment is likely to earn as a rate of return on capital over its whole life is discounted cash flow (DCF) analysis. DCF analysis involves budgeting all the expected flows of benefits in and costs out when they are expected to occur, and adjusting these flows back to equivalent present value.

The DCF approach is explained below using an example.

> Suppose that an investment has the expected flows of cash in current (real) dollars as shown in Table 4.1.
>
> This four-year project has a net value equivalent to \$794 in present-day terms, when the opportunity earnings of the capital invested in this use is 5% and thus a discount rate of 5% is used. Having a net value of \$794 in present value, after allowing for the 5% opportunity earnings forgone, means the project is earning more than 5% return.
>
> What does it mean to say, 'This project has an NPV of \$794 at 5% discount rate'? The NPV is the value of the net benefits from the investment after allowing for, say, having to pay 5% p.a. interest if all the capital were to be borrowed, or if all the capital used could have been used some other way and would have earned 5% p.a. for the life of the investment. The positive NPV means this investment is better than the alternative in which 5% return on capital could have been earned.

**Table 4.1** Expected cash flows in current (real) dollars

| Years | 1 | 2 | 3 | 4 |
|---|---|---|---|---|
| *A* cash in | 0 | 10,000 | 10,000 | 6,000 |
| *B* cash out | 10,000 | 5,200 | 5,000 | 4,000 |
| *C* net cash flow (*A* – *B*) | –10,000 | 4,800 | 5,000 | 2,000 |
| *D* net present value factor at 5% | 0.9524 | 0.9070 | 0.8638 | 0.8227 |
| *E* present values (*C* × *D*) | –9,524 | 4,353 | 4,319 | 1,645 |
| *F* net present value (sum of *E*) | 794 | | | |

The $794 isn't the actual bundle of dollar notes the investor will get to hold – it is the present-day equivalent worth of all the benefits minus all the costs of the investment that will occur over the whole life of the project. An interpretation is that, if the investor had to borrow all the funds at the discount rate of 5% p.a., then he or she could do so and pay back all the borrowings plus annual interest, and be better off by the $794 in today's dollar values. Or if the investor had all the capital and could have put it in a use that would have earned 5% p.a. for the same life of project, and instead used it in this way, then he or she has received all of their capital back plus the equivalent in present dollars of a $794 addition to the wealth they had at the start of the project, and has earned more than 5% p.a. on the capital invested.

When presented with the results of investment analysis, such as 'this investment has an NPV of $1m at 10% discount rate', the standard response is often: 'And what might that mean?' It means: $1m is the total value in today's terms of all the benefits minus all the costs of the investment, after allowing for the fact that the capital invested could have earned 10% in alternative investments of similar riskiness. The $1m is the amount that the investment will add to the investor's wealth, expressed in terms of today's dollars.

### *Internal rate of return*

However, saying that the investment in the above example will add the equivalent of $794 in today's terms to the investor's wealth after allowing for a 5% opportunity cost of capital still doesn't answer the question, 'I know the investment is earning at least 5% p.a., but what actual annual rate of return is the capital invested earning?' The rate of return on the capital invested in a project over the whole life of an investment is called the internal rate of return (IRR). With this example, the effect of using a discount rate higher than 5% is tested.

It turns out that the annual rate of return – the discount rate – that makes the NPV equal to zero is 10%. The IRR is the discount rate that leaves zero NPV. This demonstrates that the actual annual earning rate of the capital is 10% p.a. An NPV of zero means that the project has earned a rate of return over its life equal to the discount rate used.

### *Decision rule*

The decision rule is to accept the project with the highest NPV. From a range of alternative uses of a certain amount of funds involving different patterns of cash flow over the same time periods, for any particular discount rate used, the use that gives the highest NPV is the one that is most profitable. If the NPV of the project, after discounting, is positive, then the investment being analysed is better than the alternative investment (opportunity cost). For the IRR, the decision rule is to accept a project if the IRR exceeds the opportunity cost (discount rate).

### *Discount rate*

A key question is: 'In the discounting formula $1/((1 + r)^n)$, what determines the size of "*r*", the discount rate?' The discount rate '*r*' is the amount by which a benefit or cost to be received in the future is considered to be worth less than the same amount in the hand in the present. The reason for '*r*' being a positive number is that people prefer to receive a benefit now, rather than later. This is called people's rate of time preference. An indicator of how much people prefer a present benefit over a future benefit is the amount they have to be paid for them to invest a sum of money now for someone else to use and benefit from, instead of using it themselves for their own benefit immediately. For example, the 10-year government bond rate indicates how much people prefer present consumption over deferred, but low-risk, future consumption. This rate shows how much people have to be 'bribed' by a larger sum in the future (the invested capital plus interest) to persuade them to defer the benefits of consuming their capital now.

The second explanation why future benefits and costs are worth less than present benefits and costs, and thus why '*r*' is positive, is to do with the fact that when resources are invested in one use, the chance of using them in another way is gone. This opportunity cost means that the opportunity of earning net benefits from using funds in one way, say for investment *A* instead of another way, say for investment *B*, is an opportunity given up. If investment *A* has positive net benefits after allowing for the cost of the potential net benefits given up by not using the capital elsewhere, then this means investment *A* is superior to the other alternatives that are being forgone.

The choice of discount rate has some implications for the eventual choice of project. The higher the discount rate used, the lower is the value of future benefits and costs, and the lower is the NPV. So, when comparing alternative investments, high discount rates tend to rule out investments whose returns are far into the future. On the other hand, low discount rates imply that not a great deal is being forgone by not receiving the returns until later. Hence, projects with distant future returns have a better show of being acceptable if low discount rates apply – that is, they will have a higher NPV than would be the case if a higher discount rate were applied.

### *Reinvestment*

The discounting procedure has the implicit assumption that the extra funds generated through the life of the investment earn at the rate used as the discount rate for the life of the project. This assumption means that the IRR is sometimes a misleading indicator of profitability. If the IRR is markedly higher than the going market rates of earnings on investment possibilities in the economy, then it isn't a sensible estimate of the earning rate of the project. This is because the discounted cash flow method has the assumption that surpluses generated by the project over the life of the project will earn at the IRR rate for the life of the project. That is, to get an IRR of, say, 50% for the life of the project, surplus funds generated during the life of the project also have to earn at 50% until the project finishes. But if the going market rate of earning on other investments is 5–20% p.a. on capital, the needed 50% return from elsewhere (as required by the reinvestment assumption) won't be able to be earned, and thus the 50% over the life of this project won't be able to be earned either. In such a case, the NPV at realistic market opportunity earning rates gives a better indication of which project is best.

### *Benefit-cost ratio*

Another measure of the profitability of an investment is the benefit-cost ratio (B:C ratio), which is the discounted present value of benefits and the discounted present value of costs expressed as a ratio. If this ratio is greater than 1, at the desired rate of return (discount rate), the project is earning more than the required rate of return. The B:C ratio is commonly used in social benefit-cost analysis, which is investment analysis carried out from the perspective of using public resources. The magnitude of the B:C ratios of alternative projects can be compared at the same discounted rate. The project with the largest B:C ratio at the required rate of return – a B:C ratio greater than 1, of course – is the most beneficial.

These measures of the expected performance of a project – NPV, IRRs and B:C ratios – can create a false impression of accuracy. The NPV, IRRs and B:C ratios are only as accurate as the estimates of the flows of benefits and costs, and the validity of the assumptions used in the analysis and underlying the technique.

However, estimates of profitability are not the only considerations in a decision. They are part of the information used in making a decision. As well as estimating the NPV of an investment and evaluating risk, financial cash flow analysis has to be done. Other factors, often unable to be measured, also play a part in the ultimate decision.

### *Life of project and salvage value*

Estimating benefits and costs over the life of the investment requires that the relevant time scale used is the life of the project, or the planning horizon of the investor. As well, an estimated salvage value of the initial capital invested plus the capital invested through the life of the project goes into the final year's benefits. This is capital recouped. The difference between initial capital invested and capital recouped is the depreciation cost over the life of the project.

### *Inflation and the numbers*

Some explanation of the economic phenomenon called inflation is required. In practice, sums of money received at different times have different values, for two reasons: (a) the effects of inflation; and (b) the ability of investment of sums of money to earn profits and grow to bigger sums. The effect of inflation on the value of money is shown by the fact that items in Australia that could be bought for about $1 in 1967 cost five times that 25 years later. Inflation is the rise of product prices, and this causes the purchasing power of the currency to be less. A 1960s dollar cannot be compared to a 1990s dollar, because they have different purchasing powers. The point about discounting methods, though, is that even if there was no inflation, money in the hand today would be worth more than a dollar in the future because today's dollar can be invested and can grow to be worth more than a dollar in the future.

If investors expect inflation to occur, then market rates of interest reflect this expectation. Lenders will require an inflation premium to be added to the interest rate they charge borrowers. Market interest rates can be defined as:

$$m = r + f + rf$$

Where m = market or nominal rates of interest
$r$ = real real of interest
$f$ = rate of inflation expected

Thus, if 5% inflation is expected and 5% real return is required, a lender will charge approximately 10% nominal interest to a borrower.

> Suppose that a lender loans $100 for one year. Inflation of 5% is expected over the coming year. The lender wants to earn 5% real return (return above inflation) on the funds that have been lent. The lender will want $5 to cover the effect of the expected 5% inflation. A sum of $105 at the end of the year will buy as much goods and services as $100 at the start of the year. As well, the lender will want $5 more as a real return. This means the lender will require $110 for lending the $100 for one year. As well, to maintain the purchasing power of the 5% real return, the lender will want 5% inflation added to that part of the return as well. This is the '*rf*' part of the market interest rate formula. This will be another $0.25. The total return to the lender will be $110.25, an interest rate of 10.25% (0.05 + 0.05 + (0.05 × 0.05). The lender gets 5% real gain from the deal.

If the actual rate of inflation over the life of an investment exceeds the expected rate on which market rates were formed, then the investor has earned a lower real return than anticipated when making the investment decision. When this happens, borrowers on fixed interest loans gain, and lenders lose. Borrowers repay their principal and interest with cheaper dollars. During unexpectedly high periods of inflation, real interest rates can turn out to be negative. However, if inflation turns out to be less than expected, lenders get a higher real return on their lending than they had planned. Then the real cost of borrowing to fixed interest borrowers is higher than hoped for, and they lose out.

If inflation exceeds the market rate of interest, then negative real returns to the fixed interest lender will result. When interest rates reflect inflationary beliefs, then only unexpected changes in the inflation rate will benefit a borrower or a lender. This depends on which way the inflation rate goes. In competitive markets, changes in the expected rates of inflation are built into changes in the nominal rate of interest. The mechanism by which this happens is called arbitrage.

Arbitrage is the process whereby people with an eye to the main chance buy and sell, or invest and speculate, to take advantage of perceived discrepancies between markets. In this way, any gap between the ruling rates of interest and current expectations about future rates of inflation is acted on. Thus market rates of interest tend to change. This reflects the change in the beliefs about inflation held by the market participants. In practice, barriers to such market adjustments mean that it isn't safe to assume that, with inflation, market rates of interest will correctly incorporate inflation, or that they will do so quickly enough to ensure that a real rate of earning is maintained (in the short run).

Judging alternative projects to determine which is better or best involves choosing the numbers to use in the budgets. If inflation is anticipated over the life of the proposed change, some allowance can be made for this in the economic evaluation and financial planning of the change. One method is to use current (real) values for all costs and prices, rates of return and the discount rate. However, if the economic analysis is done in real dollars, and a project is chosen from among the alternatives, then the financial or cash flow budgets need to be expressed in nominal (inflated) dollars.

Financial budgets must be in nominal terms, otherwise they would understate the amount of nominal dollars needed to fund the project in future. Either current dollars and real discount rates can be used for the analysis, or all nominal dollars and discount rates – the NPV is the same. A good case can be made for doing all the analysis, both economic and financial, in nominal terms, because in either case the estimate of tax payable has to be done in nominal terms. Taxation is levied on nominal taxable income.

Using all real dollars and real rates of interest to compare projects is making the implicit assumption that the real costs and prices don't change relative to each other over the life of the project. When projecting current values for costs and receipts forwards, it is assumed that the current relationships between product prices and income and input costs remain the same in the future as they are now. If this assumption is invalid, which is possible given past inflation of farm costs compared with prices received and productivity gains achieved, then the analysis can be done in appropriately adjusted real dollar costs and prices. Another adjustment to consider is a possibility of technological gains over time reducing some costs.

A point to remember is that if an expected return on capital is calculated in real dollars, then the return is a real return. It isn't to be compared with market (nominal) rates of interest as the opportunity cost of this use of the capital. This is because the market rates of interest are made up of the percentage real return that investors hope to get, the expected rate of inflation and an allowance for risk. Only if the return on capital is calculated using nominal dollars is the return on capital comparable with the market rates of interest for uses of funds with similar riskiness.

As ever, there is a little bit more to the story of inflation and investment analysis. Our discussion of real and nominal dollars and interest rates is simplified. There is a chance that the present capital value of a productive asset itself has some allowance for expected future rates of inflation. This means, in practice, that attempts to put all values into real terms, so as to derive measures of the relative fruitfulness of uses of resources, remain slightly ambiguous. This is true if the capital values we are talking about also include a part that can be put down to beliefs about future rates of inflation. While it is best to do all the sums in either constant or future inflated values, it may be that capital values used are still something of a hybrid of real and nominal values because expectations about future inflation are built into current capital valuations.

### *Risk*

When comparing alternative investments, both risk and return are important. When considering NPV and IRR, the risk aspects have to be given equal priority with the consideration of returns. How has risk been allowed for in the analysis? Is the IRR and NPV a 'most likely' estimate? Are there also estimates of NPV and IRR for other scenarios, such as the 'Worst', 'Poor', 'Good' and 'Best' cases? How sensitive are the predicted outcomes to changes in key variables? What are the minimum or 'breakeven' levels of the critical variables at which the investment still passes the minimum criteria for acceptance? What is the probability of the breakeven levels of critical parameters happening? Once the initial discounted cash flow analysis is set up on the spreadsheet, this is when the real analysis of the decision starts. The budget has to be 'worked over' to provide information about the risk angles to enable a better-informed decision.

Implicit in the decision rule that says one should choose the project with the highest NPV is the notion that differences in risks between alternative investments have been accounted for in the calculation. This could have been done by allowing for risk in the numbers chosen, such as using expected values of numbers, including contingency allowances, or risk-adjusted discount rates, or conservative estimates of yields and prices, or conservative estimates of the life of the project and salvage values. Adding a risk premium to the discount rate is a common approach. However, this implies that risk increases exponentially into the future. Certainly, the further into the future are any projections, the more risky and uncertain are estimates about levels of project performance.

### *Investments involving different lengths of life*

When comparing investments with different lengths of life, consideration has to be given to the earning of the capital invested in the project with the shorter life, once that project is finished. In some cases, such as one of a cyclical nature – a forest plantation, say – an implicit assumption is made that the cycle of the shorter investment is repeated. In such cases, it is convenient to convert the NPV of the project with the shorter life into an annuity – an equivalent annual sum – and to compare this annuity with the annuity from the project with the longer life. In another case it may be necessary to define the next project, such as compare returns from two five-year investments with one 10-year investment. Converting an NPV to an annuity is the same as estimating the annuity that will repay the principal and interest on an amortised loan. (Use Table D in the Appendix.)

### *Investments involving different amounts of capital*

In theory, different amounts of capital being invested in projects that are being compared should not matter. If there were no constraints on access to capital, investors would simply choose the most profitable investment, then the next and the next, and so on. In assessing alternative investments involving quite different quantities of capital measures, all the measures of efficiency (return on capital, IRR and B:C) and absolute size (NPV and addition to wealth) have to be considered.

If an investment passes tests of economic efficiency to do with required rates of return on capital, then the second criterion for judging it is the financial feasibility of the project. This involves projecting the nominal cash flows as they are expected to actually occur in the future. This differs from the economic analysis in that actual flows of cash only are included; thus salvage values of capital items at the end of a project's life are counted in the economic analysis, but are not included in the financial analysis unless the items are actually going to be cashed in. Financial analysis has to be done in the nominal (inflated) dollar values of the relevant future time when the cash flows occur. Financial analysis is useful to indicate the size and timing of cash deficits, and how long it might take to recoup the cash outlays and make some profit. Testing this involves constructing a nominal cash flow budget of the expected benefits and costs, charging a market rate of interest annually on cumulative cash deficits, and earning interest annually on cumulative cash surpluses. The cumulative net cash flow will indicate the size of the peak debt and the time before the project generates a cumulative cash surplus. This information can be used to establish appropriate financing arrangements for the project. The financing decision is about how much money needs to be raised, in what ways, and under what terms. The financial markets provide the means by which an investor may get access to various forms of finance.

## Discounted cash flow analysis

A farm family is considering improving some pasture on their 2,000-hectare property. They have a 400-hectare area of arable land of quite uniform soil characteristics, and mostly flat to gently sloping, that has not been fertilised or sown with introduced pasture and is currently low-producing native pasture. Details of the proposal are as follows:

- The 400 hectares currently carries six DSE/ha p.a. This is three breeding units (ewe, lamb for five months, 1/40th of a ram).
- The unimproved pasture currently receives no annual fertiliser.
- The expected GM/DSE from the current stock is $30/DSE. Extra stock carried are expected to produce a GM/DSE of $30/DSE.
- The pasture establishment costs $250/ha. This is for seed, sprays, initial heavy fertiliser application and cultivation.
- From year 2 there will be an annual pasture maintenance cost of $40/ha made up of annual fertiliser topdressing and some pasture topping and weed control measures.
- To allow for risk, a supplementary feed cost of $5/DSE p.a. is included for every extra DSE carried.
- Extra casual labour will be employed at $2,000 p.a. for the life of the project.
- A breeding unit comprises a ewe and 1/40th of a ram, and is equivalent to two DSEs and has a capital value of $100.
- The extra breeding units carried have a salvage value of $100 ($50/DSE).
- The capital invested in pasture establishment has a salvage value in year 10 of 50% of the initial capital invested.

The method of financing will affect tax paid, as the interest is a tax-deductible expense. As well, the capital expenditures on the pasture development are deducted for tax purposes at 10% p.a. To allow for these effects, and for the farmer utilising tax averaging, a simple method is used to make an approximate estimate of the tax effects. Tax effects won't be zero, and including an approximation of these effects is better than doing the analysis as if there will be no tax effects. In this case, as the analysis is in real dollars, extra income tax in real terms in this development is approximated as being equivalent to paying $0.10 for every extra dollar of the real annual net cash flow.

Because the proposal involves streams of cash in and cash out over a number of years, DCF analysis is necessary (see Table 4.2). As the proposal involves adding to existing investment in land and stock and other farm capital, a partial development budget is required. That is, only the extra costs and extra returns involved are included in the analysis.

From Table 4.2, this proposal looks economically viable as the expected returns after allowing for risk compare favourably with the after-tax weighted average cost of the funds of 10% real return. The most likely (expected to occur four to five times in 10 years) GM/DSE of $30 gives an NPV of $93,000 at 10% real discount rate, and an IRR of 20% real. A GM/DSE of $25 gives an NPV of $33,000 at 10% discount rate and an IRR of 14% real. A GM/DSE of $20 gives an NPV of $0 at 7% discount rate – that is, an IRR of 7% real. The minimum acceptable real IRR is 10%. This would occur if the GM/DSE was $23 p.a. This outcome could be reasonably expected to occur two to three times in the 10 years of the planning period. Overall, the likely return on capital looks sufficiently greater than the opportunity cost for the project to go ahead. Further scenario testing can be done. For example, how would it look if the relatively poorer GM/DSE occurred in the first three years? How does a serious drought in years 2 and 3 affect the overall result?

Having done the economic analysis in real terms, and finding that the proposal looks attractive, the financial analysis is done. This involves estimating the expected cash flows that may occur once the expected 3% p.a. inflation is included. Note that in doing the financial analysis, the actual sales and purchases of animals need to be included when they occur, not the gross margin/DSE figure. As well, the salvage value of initial capital isn't included, as the initial capital invested in pasture and most of the animals isn't actually cashed in at the end of this project.

**Table 4.2** Discounted cash flow budget

| Year | 1 | 2 | 3 | 4 | 5 | 6 | 7 | 8 | 9 | 10 |
|---|---|---|---|---|---|---|---|---|---|---|
| Total DSE/ha | 4 | 10 | 12 | 15 | 15 | 15 | 15 | 15 | 12 | 10 |
| Extra DSE carried (more than unimproved state) | −2 | 4 | 6 | 9 | 9 | 9 | 9 | 9 | 6 | 4 |
| **Benefits** | | | | | | | | | | |
| Extra Gross Margin | | 48 000 | 72 000 | 108 000 | 108 000 | 108 000 | 108 000 | 108 000 | 72 000 | 48 000 |
| Capital salvaged-livestock | 40 000 | | | | | | | | 60 000 | 40 000 |
| Capital salvaged-pasture improvement (50%) | | | | | | | | | | 50 000 |
| Total Benefits | 40 000 | 48 000 | 72 000 | 108 000 | 108 000 | 108 000 | 108 000 | 108 000 | 132 000 | 138 000 |
| **Costs** | | | | | | | | | | |
| Net income forgone | 48 000 | | | | | | | | | |
| Pasture establishment costs | 100 000 | | | | | | | | | |
| Livestock capital | | 80 000 | 40 000 | 60 000 | | | | | | |
| Annual pasture maintenance | | 16 000 | 16 000 | 16 000 | 16 000 | 16 000 | 16 000 | 16 000 | 16 000 | 16 000 |
| Supplementary feed | | 8 000 | 12 000 | 18 000 | 18 000 | 18 000 | 18 000 | 18 000 | 12 000 | 8 000 |
| Casual labour | 2 000 | 2 000 | 2 000 | 2 000 | 2 000 | 2 000 | 2 000 | 2 000 | 2 000 | 2 000 |
| Total Costs | 150 000 | 106 000 | 70 000 | 96 000 | 36 000 | 36 000 | 36 000 | 36 000 | 30 000 | 26 000 |
| Net Benefits before tax | −110 000 | −58 000 | 2 000 | 12 000 | 72 000 | 72 000 | 72 000 | 72 000 | 102 000 | 112 000 |
| Net Benefits after tax | −99 000 | −52 200 | 1 800 | 10 800 | 64 800 | 64 800 | 64 800 | 64 800 | 91 800 | 100 800 |
| Net Present Value 10% | 93 679 | | | | | | | | | |
| Net Present Value 20% | 0 | | | | | | | | | |
| Nominal NCF after tax with 3% inflation p.a. | −101 970 | −55 379 | 1 967 | 12 155 | 75 121 | 77 375 | 79 696 | 82 087 | 119 778 | 135 467 |
| NPV at 13.3% discount rate | 93 679 | | | | | | | | | |
| Cumulative NCF after tax with 3% inflation p.a. | −101 970 | −157 349 | −155 382 | −143 227 | −68 106 | 9 269 | 88 965 | 171 052 | 290 830 | 376 296 |

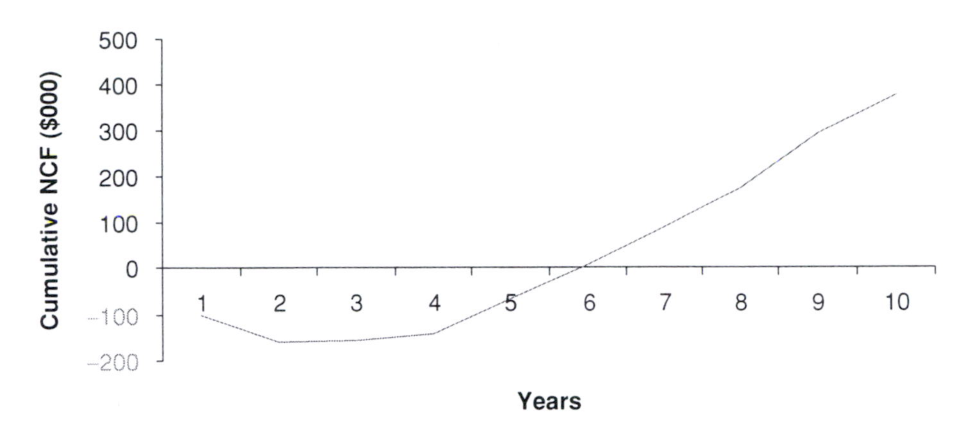


**Figure 4.1** Financial analysis: Cumulative net cash flow after tax with 3% inflation p.a.

The size and timing of peak debt is as shown in Figure 4.1: peak debt of $157,000 in year 2, and cumulative net cash flow becomes positive in year 6. Various loan options and repayment schedules would be investigated to see if the project, which is economically viable, is also financially viable.

## Sound judgment and perspective

To provide a sensible answer to an investment question, an analyst needs to bring to bear on the question a balance of appropriate knowledge of disciplines and their linkages and the ability to see the question in the appropriate perspective(s). The theoretical foundations, both technical and economic, have to be sound and the alternative futures soundly conceived and well defined. Further, generally, only a few of the numbers in the analysis really matter. Finally, the analyst should recognise well what is known and not known about the futures he or she is investigating, and heed the observation of a major contributor to the development of cost-benefit analysis, E. J. Mishan:

> As several conscientious economists have pointed out, the outcome of all too many cost-benefit studies follows that of the classic recipe for making horse and rabbit stew on a strictly fifty-fifty basis, one horse to one rabbit. No matter how carefully the scientific rabbit is chosen, the flavour of the resulting stew is sure to be swamped by the horseflesh. (Mishan 1982, p. 149)

Mishan's dictum about horse and rabbit stew remains a pertinent, potent reminder to benefit-cost analysts about the folly of thinking that more and more elaborate analysis can somehow change the uncertain aspects of an investment into certainties, or convert the unknowable into the knowable. Sophisticated, rigorous thinking underlying some simple sums about how the world may be remains a fine recipe.

## EVALUATING INVESTMENT IN FARMLAND

At some time in their career, most farmers – and many business people – will buy farmland. The decision to buy farmland is the biggest single business decision a farmer ever makes. 'When to buy more land? Could it possibly be worth that much to us, the farm family? How much should we, could we and would we pay for this land?' These are the searching questions that farm buyers and their families must answer before they commit themselves to a large, long-term investment in farmland.

The land market is remarkably complex. First, 'they don't make it any more'. Land isn't a mobile resource. Land is durable; it varies greatly in type, quality and uses; and importantly, people often tend to regard land as more than simply a factor of production. Thus the word 'value', with respect to land, has many meanings in everyday farming language. Despite this, all landowners have some reserve price that is the minimum price they would be prepared to accept for their land. Similarly, all potential land buyers have some offer price that is the maximum price they would be prepared to pay for a piece of land. The main economic influence on offer and reserve prices is likely to be the expected net benefits, especially profitability, of owning agricultural land. This means that the offer and acceptance prices of prospective and present farmers depend considerably on their respective expectations about the net income from the production of agricultural commodities from the land, the expected change in net return from using the land, and any change in capital value expected from future sale of the land. Such expectations, and attitudes to risk, are major determinants of the offer and acceptance price. Other factors that underlie the respective prices include present financial position, availability of finance, taxation, the available alternatives to buying or selling land, and the non-pecuniary aspects of owning farmland.

Land is also purchased for its potential or its security, rather than for its immediate productivity. Buyers see scope for increasing output and income by implementing technological advances. If technological progress over time lifts the potential productivity and profitability of farm production, then this will end up being reflected in land values. Note, though, that real land values can fall or rise. The main point in discussions about land values is that whatever happens in the way of profitability of production, be it a sustained rise in product prices because of increased demand from a new market, or protection and assistance granted to farmers by governments, or new technological possibilities, the expected future net gains all end up in the present value of the land.

In essence, land values reflect the incomes and capital gains that are expected from the management and control of the land, after the potential purchaser has made some allowance for the uncertainty about prices and yields in the future. The pressure of urban demand for land close to cities or for recreation and hobby farms is an important consideration in some sectors of the agricultural land market. Nowadays, the demand for use for recreation and country living accounts for an increasing proportion of small farm sales in favourable areas of Australia. Also, the rise in capital value of many farmers' land has given them greatly increased borrowing power to expand their holdings; this partly accounts for a good deal of the demand in the land market.

## Purchasing a farm

Farm buyers have different motives for buying. Some of the common ones are: way of life, profits, capital gain, hedge against inflation, security, speculation and recreation. It is impossible to give simple recipes about purchasing a farm. This is because the value that a prospective buyer places on a farm incorporates a wide range of matters, only some of which are economic factors that can sensibly be given dollar values. Before going ahead with farm expansion plans, it is worthwhile to have a careful look at the potential for further developing the existing farm. In some cases, many of the benefits that are expected to flow from farm expansion can be achieved by developing or diversifying the existing holding. Farmers often find that until their own farm is reasonably well developed, an investment in developing the existing landholding under good management will bring greater returns than buying more land. Each case has to be judged on its particular merits because there are so many variables to consider.

Once the decision has been made to investigate buying farmland, the first aspect to consider is the area where the farm for sale is situated. Features to investigate, from as many informed sources as time permits, are: history of settlement, climate, proneness to natural disasters (drought, flood, fire), present land use and agricultural production systems, crop yields and stock carrying capacities, recent and forecast trends in the industry, educational facilities, employment opportunities, and the expected influence of urban growth, mining developments and national parks. When looking at a particular farm, the number of times it was sold in recent years, access, location, and the state of the house, structures and improvements will obviously determine the appeal it has for the purchaser and his or her family. A major point to consider here is how much capital needs to be invested to bring the structures up to a standard that suits the would-be purchaser. Any analysis of a farm must consider both the present situation and the potential for change and development.

## Valuing farmland

There are a number of ways of calculating how much to offer for a piece of farmland. In essence, these methods condense to how much a bidder:

- should pay, according to strictly economic criteria;
- has seen others pay for similar land and might have to pay for the land in question (called market value);
- could pay, according to their ability to finance the deal; and
- is prepared to pay, all things considered.

### *What should be paid*

The 'should pay' economic value of the investment is the maximum amount the buyer can justify paying in order to earn a minimum rate of return that is required by the investor from the investment. This land price is based on the expected after-tax returns on the investment, made up of annual operating returns and expected capital gains, if any. The way to calculate how much an investor should pay is to use the standard discounted cash flow method of investment analysis. The essence of this approach is the notion that the value of an asset depends on the future streams of net benefits that the asset is expected to generate and the rate of return on the investment that is required.

> Suppose an asset promises to provide a net return of $100 p.a. If a buyer wants to earn 10% on his investments, then he could pay $1,000 for the asset, and the $100 earned would represent the 10% required return on capital. This is called the income capitalisation method of valuing an asset, and is essentially the approach used in valuing farmland according to expected net earnings. However, complications such as inflation and subsequent capital investment, interactions with the existing farm business, and sometimes capital gains from increases in land

> value, mean that the more complicated method of investment analysis, called discounted cash flow analysis, has to be used to properly value farmland.

To work out how much a parcel of farmland is worth to a particular business, first a planning horizon is defined – for example, 10–15 years. The purchase of farmland by private investors must be analysed over the planning horizon of the investor, whose future intentions and, in particular, attitudes to the risk, uncertainty and other unknowables about the future are critical elements of the investment decision. Commonly, farmers buying land use from 10 to 15 years as their 'life of project' or 'planning horizon'. With proper consideration of salvage value at the end of any planning horizon, the value of the investment remains the same regardless of the life of project used in the analysis, because the assumption is made that the land is salvaged at a value based on the next buyer valuing the land according to similar expected net returns on the land capital in the future.

To define the maximum offer price in order to get a minimum return on capital, the expected annual net earnings from the asset need to be estimated. These estimates form the basis for a DCF analysis. As well, for the analysis, the asset is purchased at the start and salvaged at the end of the planning period using an assumed price of the land. The internal rate of return is calculated. If the IRR equals the required rate of return, the investor has paid a price for the land that earns the required rate. If the IRR is less than the required rate of return, the investor would have to acquire that piece of land for less than the price assumed, for it to earn at the required rate. If the IRR is more than the required rate of return, then the investor would be able to pay more than the assumed price for the land and still earn the required rate. Using a spreadsheet, it is straightforward to change the purchase price and salvage value until the IRR equals the required rate of return for the land. This defines the maximum the land buyer should pay on the basis of expected economic net benefits. If inflation is expected, then the value of land has to be estimated using nominal values.

The value derived using DCF methods is treated as giving one 'fixed point' for the potential buyer to focus on. It isn't the 'correct' value of the land, because there are a number of factors not yet accounted for, and there is always uncertainty associated with the figures used. However, a range of land values and related returns on the investment can be estimated in this way. The aim is to construct a profile of possible values and returns on capital for possible scenarios, based on the 'measureable' elements of the decision. This profile then acts as a basis for adjustments for the important other factors involved in the decision.

The goals and objectives of farm families are the key to the land valuation question. To a considerable extent, the land purchase decision is one that involves long-term objectives, such as the desire to continue farming because that is what farmers love, know and do best. Then there is the wish to grow to an adequate size to enable heirs to continue farming in the future, and also to build up an adequate-sized 'superannuation' for future retirement. Thus, short-term alternative rates of earnings from a similar-sized investment may never come into the decision. It is little wonder that the apparent capitalisation rates used by farmers often seem low. Anomalies are found. For instance, rising land values have been found coinciding with apparent falling net returns per hectare. This can be explained partly by expectations about inflation and partly by changes in the real cost of borrowed money. The availability of credit to farm buyers affects their bidding power. Thus a more accurate relationship explaining land values might be the relationship between not just expected net cash flow, land value and the implied capitalisation rate, but between net cash flow plus borrowings and land value and the capitalisation rate. The size of the parcel of land is also important, with smaller blocks of land selling for more per hectare than larger areas.

Finally, there still remain the benefits of land ownership that are difficult to measure. These can include the fulfilment of some deep-seated intrinsic wants, or there could be the more practical advantage of proximity. As most farmland is sold in parcels to neighbours, the effect of proximity and competition between neighbours on purchase prices can be marked. Another difficult-to-assess attribute of land value is farmers' preference for longer-term wealth over short- to medium-term operating profits. To a considerable degree, it is likely that farmers are motivated as much or more by maximising their wealth or net worth, as they are by motives of achieving the most short-term profit. Among other reasons, wealth, and thus equity, is valued because it reduces risk by reducing gearing and increasing liquidity.

So far, we have discussed deriving some economic values for what a buyer, in theory, should pay for a parcel of land – that is, when to stop bidding in order to have a good chance of earning the required rate of return. This is useful information for a starting point. However, more practical information is also needed. Two particularly practical bits of information are: 'What is the current market value of similar properties in the area, as suggested by the prices other buyers have recently paid for farmland that is similar and is for a similar use?' And,

'How much could I pay for the land given the relevant financial constraints?' In practice, these two considerations – current market prices and what can be financed – are the key bits of information in decisions about land value.

### What others pay: Market price

Recent prices paid per hectare for similar farmland that is used in similar forms of production to the systems the intending buyer will employ are an indicator of current land values. However, all parcels of land are unique. Great care is needed to allow properly for important differences between the market's prices and the characteristics of sales of particular parcels of farmland – for example, housing, improvements, topography, slope, aspect, history, buyer's and seller's motives, vendor finance terms, and so on.

### What could be paid

How much a buyer 'could pay' for farmland refers to the financial feasibility of the investment. Whereas the 'should pay' criterion defined a maximum offer price that is consistent with the minimum acceptable return on capital, 'could pay' defines the maximum offer price according to how much capital – equity and debt – is available and can be serviced. The financial feasibility of investment in farmland, at various possible prices for the land, is one of the keys to appraising the land price a particular potential buyer could offer. In any year the farm family is concerned with generating sufficient cash to meet operating expenses, living expenses, replacement costs, taxes, and interest and principal payments. The annual cash surplus after all the annual cash sources have been accounted for and have been met, but before new interest and principal payments, indicates how much extra debt from purchasing the land could be serviced, for various types of loan and terms of loan servicing.

To estimate financial feasibility, first expected annual cash surplus from production, before new debt, is estimated. This expected annual cash surplus is the amount available to meet annual interest and principal repayments. For an amortised loan, the expected annual cash surplus represents the annuity available to meet the amortisation payment (interest and principal).

> One hundred hectares of land is under consideration. The expected annual cash surplus from all sources and after all cash uses, but before new debt servicing, is $21,000. To be on the safe side, say, as financiers are prone to doing, that 70% of this sum can be confidently expected

to be available in any year – say $14,000 cash surplus is expected to be available to service debt every year. This sum is a nominal value and is expected to increase at the rate of inflation over time. The debt terms are calculated using nominal, not real, interest costs. The amount of an amortised loan that can be serviced by this sum is given by the present value of an annuity (Appendix C). With no initial deposit, $14,000 could service a 10-year loan, at 15% interest, of $14,000 × PV Factor 5.0188 = $70,263. As a check on this, $70,263 × Annuity Factor (Appendix D), 10 years, 15%, 0.1992 = $13,997 annuity. The buyer could offer up to $70,000, or $700/ha, for the land. When equity capital is available, the financial feasibility of an investment is assessed as follows:

Maximum offer price = Equity capital + [Cash surplus available to service debt × Present value of annuity]

The annuity is calculated as: $[(1 + i)^n - 1] / [i(1 + i)^n]$,

where $i$ = interest rate
$n$ = number of years

If $200/ha is available in equity capital, then in the example above $900/ha could be paid for the land.

A spreadsheet is useful for quickly devising a schedule of maximum financial feasible price per hectare with various interest rates, loan maturities, annual cash surpluses and equity capital available.

### *What the buyer is prepared to pay, all things considered*

In determining how much to actually pay, after estimating the economic 'should pay', considering market values and calculating the financial 'could pay', the potential buyer knows, at the very least, how much *not* to pay. How much a buyer is prepared to pay involves consideration of market values of similar land, and of factors not easily amenable to definite dollar values but that are still important. The case studies that follow indicate some of the considerations that various potential buyers of farmland included in their evaluations.

## Farmland purchase case studies

Several farm purchase decisions were analysed by applying farm management techniques – namely, a 'first-look' economic and financial appraisal, followed by a more thorough discounted cash flow investment analysis using a discounted cash-flow land price purchase appraisal model (Madden and Malcolm 1996).

### *A land price purchase appraisal model: Method and assumptions*

The method used is the standard discounted cash flow technique, with the question of land price to be solved for a defined required rate of return. That is, the land price that gives an NPV of zero at the required rate of return is the maximum offer price in terms of economic criteria. This calculation can be done for two situations. First, for the situation where the perfect capital market assumption holds, this means that borrowing and lending rates are the same and the investor can borrow and lend the sums required over time to generate the pattern of net cash flow that matches consumption requirements and maximises utility. In this case, the source of investment funds doesn't matter and the tax effects of borrowed finds are not accounted for. This analysis gives the IRR associated with a given land price which is return on total capital invested and a measure of the efficiency of total capital.

Second, the maximum offer price can be determined for the situation where both equity and debt capital are used, where borrowing and lending rates are different, and where constraints on borrowing and debt servicing ability exist. This calculation estimates, for a given debt-equity ratio and given costs of debt, the return on equity after the tax advantages of interest deductibility are accounted for.

The assumptions and premises of the land price model are:

- Initial capital costs are incurred in year 0.
- The criterion used is nominal return after tax.
- The planning period is 15 years. Economic and financial calculations are for 15-year periods.
- Debt servicing ability is based on the minimum after-tax annual net cash flow plus any defined extra annual net cash flow available to service debt.
- Depreciation of new capital investment (addition to existing farm capital) is estimated as initial capital investment minus salvage value. Depreciation of pre-existing capital is deducted in the estimate of annual farm operating profit.
- Land is salvaged at the same real price as it is purchased. That is, the purchase price increased by the compound expected inflation rate. The rationale is that, if real farm operating profit and net return to land is unchanged, then the real value of the land won't change.

- Machinery capital is salvaged at defined rates based on expected depreciation over the planning period.
- Livestock capital is salvaged at the same real value as the initial value.
- Expected annual inflation is used to adjust all items in the budget to nominal values. Land values, prices and costs are nominal, and the key criteria are nominal after-tax return to total and equity capital.
- The tax deductibility of interest on funds borrowed to finance the investment is allowed for, because the method of financing the deal (equity:debt, interest-only loan, and so on) ultimately plays a role in determining the value of an asset.

## Beef business land purchase

This first case study involves a family running a commercial beef business. The property the family was interested in purchasing was two hours' drive from where the current business was located. The owners of the beef business were a family, with two sons in their mid-twenties. A major motivation for wanting to purchase an additional grazing property was the need for the current business to be expanded so that the sons could each have a viable farming business following the planned retirement of the parents. They were initially attracted to this district by land values that they considered were substantially less subject to non-commercial agricultural influences, compared to their current location which was affected by proximity to a major inter-city traffic route. (Ironically, later this was seen as an advantage.) The family had 90% equity in their farming business and the parents had other assets set aside to provide for their retirement. The aim was that the two sons would take over the existing beef property and a similar-sized beef property that was yet to be purchased. It was envisaged that eventually each son would own and operate the two properties as separate entities, and develop and expand these businesses according to their own needs and aspirations.

The property under consideration was 1,248 hectares in a high rainfall region, for sale by private treaty. The farm comprised a majority of fertile basalt-derived soils, with the balance made up of heavy grey clay flats, with low-fertility, shallow, acid soils on hilly areas. Expected annual rainfall was 700 millimetres. The property had a history of regular annual applications of superphosphate. Lime had already been applied to the arable land. The average Olsen P status (as explained elsewhere, a measure of available phosphorus) was 12. Average soil acidity, measured as soil pH, was 5.8. The pastures were a mix of sown perennial and annual pastures, with native grass pastures on the hills. Carrying capacity was estimated to be an average over the whole farm of around 10 dry sheep equivalent per hectare all-year round, with the carrying capacity of the flats up to 12–15 DSE per hectare, and some of the hills carrying 7–8 DSE/ha. This meant that the interested buyers considered there was still potential for improvement in carrying capacity, though this potential wasn't fully included in the analysis of an offer price. The wet and cold winters were the usual feed-limiting period, though feed supply in autumn could be just as limiting if the annual autumn break was delayed past late April. Around 70% of the total annual feed supply was produced from late August to mid-January. The property had sound improvements (cattle-handling facilities, fencing, sheds, water supply) and a well-maintained, medium-sized, brick house that was built in the 1960s. The nearest major business centre was 30 kilometres away.

The family evaluated the prospective purchase of this property. They focused on carrying capacity and asking price, converting this information to a figure of dollar cost per DSE of carrying capacity to use as a comparison with other properties of a similar type in this and other areas. They also estimated potential 'gross income' (which was estimated as cash annual sales from a steady-state herd) and deducted what they called 'operating costs' (basically, a mix of variable and overhead costs, with some annual capital development expenses also included) to derive 'operating profit'. This figure was then expressed as a percentage of total capital to give 'return on capital'. The family had decided that prospects for the beef industry were good, and they made an offer of $1,550 per hectare. The financial aspects of the purchase were assessed simply as an estimate of expected net cash surplus and the likely interest on the likely borrowings based on the asking price.

A 'parallel' analysis of the prospective farmland purchase was conducted involving a preliminary (first-look) annual steady-state whole farm analysis and a 'life-of-project' discounted cash flow analysis. The important points of these analyses are outlined below. The 1,248 hectares was expected to carry a beef-breeding herd of 1,000 head. This was calculated by converting the expected DSE per hectare of the property to total DSEs, and dividing by 12 DSE to estimate how many breeding cows could be run on the property. An expected value annual gross margin, in real dollars, for the medium term, per breeder, was estimated to be $325 (that is, based on the farmers' expectations of a 20% chance of $300 gross margin, a 60% chance of $325 gross margin and a 20% chance of $350 gross margin). Expected annual total gross margin for the farm was $325,000. Annual overhead costs were expected to be $160,000, leaving an expected annual operating profit of $165,000 before tax. This is a return on land, livestock and plant capital. If 1,000 breeders were valued at $800 per head, total livestock capital would be $800,000. Plant required was estimated at $100,000. A return of 5% nominal before tax on cattle and plant capital would leave $120,000 ($165,000 minus $45,000) expected nominal returns before tax on land. Using the income capitalisation method, and valuing the land at $2 million, or $1,600 per hectare – a figure that was in line with the typical land values in the region – the implied nominal return before tax was 6% p.a.

As a further check, DCF analysis was done. Account was taken of the timing of cash flow and including transfer costs, risk contingencies, salvage values, the amounts of equity and borrowed capital, and the tax deductibility of the proposed financing arrangements. A 15-year planning horizon was used. The farmers were planning to invest $2 million equity capital and would have ample capacity to borrow the balance required for the total investment cost. A purchase price of $1,934,400 for the land, with $2 million equity capital, and all other funds borrowed over 15 years at an interest rate of 8% p.a., was predicted to return 6.84% nominal return after tax per annum. With full equity capital invested and no borrowings, the expected nominal rate of return after tax on the investment was calculated to be 5.3%.

The required price for land was sensitive to the required rates of return. Usually farmers' attitudes to expected required rates of return are flexible over a range of returns, and the farmers in this case were no exception. Their view was that as long as the expected return on investment was within a reasonable range relative to the rates of return generally available in their type of farming, then that would be satisfactory. The expected returns in this case would have been acceptable. Small changes in expected returns on capital are associated with considerable variations in land prices.

The focus of these farmers was mainly on a medium-term view of what they believed beef production land was 'worth'. These views were influenced strongly by their optimism about the economic prospects of the beef industry (and a strong belief in 'their' industry), as well as by prices of recent sales in the district. They reasoned that they would be easily able to finance the purchase at the asking price, as using the proposed mix of equity capital and borrowings, around $65,000 was required to be serviced from the expected annual cash surpluses of $130,000 from the new property.

## Cereal property purchase

This land purchase case study involved a farmer in his fifties operating a cereal-grazing operation. He farmed 1,280 hectares with neither long-term nor medium-term debt, and had a net worth of $970,000. Soil type was mostly Mallee sandy loam with some small areas of self-mulching grey clays. Typically, 600 hectares of crop would be grown each year, using good, mostly second-hand machinery. A flexible crop rotation was used, consisting of a cereal, a grain legume and canola crops, with a pasture ley period, and with most paddocks being fallowed for either long or short periods. In most years, and depending on the season, one or two paddocks would be direct-drilled with a cereal or grain legume crop. About 1,500 sheep (1,000 wethers for wool and 500 first-cross ewes for prime lambs) were also run. These were used to graze pastures in the non-cropping phase, utilise crop stubbles and to 'clean up' fallow. The main assets of the farm business were the land, livestock and cropping plant.

This farmer had an opportunity to buy a block of land about three kilometres from the home farm. An elderly neighbour's farm was being sold, consisting of four separate blocks of land, three of which were 256 hectares in size, and one of 125 hectares. The case study farmer was interested in adding one of the 256-hectare blocks of land to his current farm. Two of the blocks of land that were for sale were located conveniently and were quite similar in terms of soil quality, cropping history and improvements. Either block of land would fit into the farmer's business well. All the land was for sale by public tender. Prior to the tender date, another neighbour had approached the case study farmer. The neighbour explained that one of the two blocks of land the case study farmer was interested in would fit into his farm system more conveniently than the other block of land. The neighbour suggested that they tender on only one of the blocks of land each. The two farmers had briefly considered forming a loose 'consortium' with another neighbour who was also a prospective buyer of some of the land, to bid for the whole farm. Then, by prior arrangement, they would each obtain the block of land they wanted. This idea was rejected as being 'too involved', and instead the case study farmer agreed with his neighbour to tender for only one block of land, while the neighbour agreed to tender for the other block of land.

In arriving at how much to offer for the block of land he wanted to buy, the case study farmer based his offer price on recent sales in the district of similar land. In the previous five years, relatively little wheat-growing land in the region had changed hands. When a sale had occurred, it had been for relatively depressed prices of around $400 per hectare. However, after some changes were made in cropping systems (principally the introduction of canola, a valuable oilseed and rotation break crop, and an intensification of crop rotations), along with recent very strong grain and oilseed markets and a recent encouraging recovery in wool and prime lamb markets, optimism among farmers in the region was growing rapidly. Thus sale prices of cropping land in the region had risen considerably. This had resulted in an increase in the number of farms coming on to the market after years of inactivity in the local land market. By 2001, typical cropping land was selling for prices ranging from $500 to $625 per hectare, depending considerably on the soil types, level of improvements, size of the block and the management history.

The case study farmer knew the neighbouring block of land well. He also mentioned a traditional rule of thumb: 'If the land can produce a tonne of wheat to the acre, then the price of a tonne of wheat is the price to pay per acre for the land.' At $200/tonne of wheat, this 'rule' gave a price of $450 to $500 per hectare. On the basis of the increased optimism and recent sale prices around at the time, and the fact that the extra 256 hectares could be farmed with the farmer's existing cropping plant, and with some small extra investment in stock and extra casual labour, the farmer settled on submitting a tender price of $525 per hectare.

Having no debt on the existing farm and a net worth of $970,000, the farmer calculated that he could borrow the full $134,400 required to buy the land plus $15,000 transaction costs and $12,500 for extra sheep, by using a collateral of 50% of the value of the new land (the bank's lending rule) and a mortgage on a portion of the home farm. The debt of $160,000, if borrowed as a 15-year amortised loan at 8%, would require around $19,000 per year to service. The farmer believed that this sum would be available from a medium-term average (as defined by the farmer) gross margin of $150 per hectare from the cropping–grazing rotation. This promised a total gross margin of around $38,000 from the block of land, minus some extra labour and other overhead costs of around $16,000. In years when the net cash flow would be insufficient to service the debt, there would be some cash surplus available from the existing business.

As it turned out, the farmer's tender offer of $525 per hectare was insufficient to buy the land. The buyer was a farmer who farmed a similar-sized operation in the local area. He paid $540 per hectare.

Subsequent discussions revealed that the main motivation behind the offer by the successful buyer of $540 per hectare was the fact that he had two sons, aged 15 and 17, both of whom were keen to go into farming. The successful bidder had a farm of 1,200 hectares, which could only support one family. The farmer felt that he needed to expand his farm so that at least one of his children could go into farming.

The unsuccessful case study farmer's situation was different. His only son, who had earlier shown significant interest in returning to the land and had attended an agricultural college, and who had been home working on the farm for six months, was in the process of deciding to pursue an alternative career. The case study farmer said, 'If my son had been definite about staying on the farm, I'd have gone a bit higher on the tender.'

A parallel analysis was done for this study. The proposed land purchase was analysed separately using the standard farm management techniques – essentially a 'first-look' analysis and DCF analysis. A preliminary analysis of the performance of 256 hectares of land in a hypothetical steady-state year was conducted.

The case study farmer had continually stressed that in many years in the region low rainfall caused poor crops, and that the variability of yields around the most likely expected yields was large. In good years, 2.5 tonnes per hectare of wheat was possible, though more often 1.5 t/ha was achieved and, too often for the farmer's peace of mind, yields not much over 1 t/ha were produced. These low yields did not cover growing costs. It was agreed that if things went well, the land could generate a most likely gross margin per rotation/hectare of $150/ha, or $38,000 total gross margin. Extra overheads of $16,000 left a most likely expected return on extra capital invested in the land and livestock of $22,000 p.a. before tax. A contingency sum of around 20% ($8,000) was then built into the analysis to allow for the risk. A return of $1,000 (8% real) on the sheep capital was estimated, based on an expected gross margin of $15/DSE, and a net margin of $10/DSE. This left a likely return on the land capital of $13,000, or 9.6%, before tax.

Using the DCF land price model, if full equity was put in to purchase the property at a purchase price of $525/ha the nominal return after tax was expected to be 8.6%. At 50% equity, this nominal after-tax return on capital increases to 9.7% because of the tax deductibility of interest. With all capital borrowed, the nominal return after tax would be 12.6%.

The analysis of this potential land investment from the viewpoint of return on marginal capital shows that extra land added to the current whole farm, with associated spreading of overheads, can result in returns on marginal capital that are higher than returns on total capital. This predicted, relatively high, rate of return was a source of wry amusement to the farmer, who believed that such returns only ever happened 'on paper'. Typically, on average, nominal returns on total capital in Mallee wheat farming range from zero (in poor years) to 3–6% in good years.

This farmer had a firm view about what could be paid for farmland in this region, based on past land prices. He had experienced 'tough times' often enough in a 30-year farming career. He was acutely aware of spectacular mistakes made by farmers in the district who paid too high a price for land and had suffered massive capital losses from their folly.

## Dairying property purchase

This land purchase case study involved a dairy farming business that currently milked 180 cows on 80 hectares of irrigated and dry-land pasture in the Goulburn Valley. The home farm had a water right of 500 megalitres that enabled 50 hectares of irrigated perennial pasture to be grown. The balance of the farm was rain-fed annual pasture. The husband and wife, aged in their early thirties, and with young children, provided all the labour.

With some strategic feeding of grain, the herd could average 200 kilograms of butterfat per cow annually. A typical annual gross margin per cow was around $660 per cow, making a farm total gross margin of $120,000. Overheads were around $80,000, leaving around $40,000 annual operating profit when things went well.

The total assets amounted to land, stock and machinery worth $800,000 (land worth $550,000, cows worth $150,000 and machinery worth $100,000). Equity was high with only a small medium-term debt to be serviced at 8% interest rate, requiring less than $10,000 to service. The family also had $300,000 of off-farm assets from an inheritance that was currently invested in the stock market.

The family had only been dairying for four years and aimed to continue for another 10–15 years. Their main aim was to achieve growth in their net worth. To do so, they considered that they would need to increase their debt and the number of cows they milked. They believed that within two to three years they could milk over 500 cows if they bought more land nearby and milked 250 cows on the extra land, using a farm manager. As well, importantly, improved grazing management would enable an extra 100 cows on the home farm, with a little extra investment in feed and infrastructure on the home farm, and with an extra employed labour unit. The first need was to buy more land. The milking facilities on the existing farm were limiting. Handling any more than 280 cows would involve too much milking time.

The farmers settled on trying to buy an 80-hectare (with a 400-megalitres water right) farm that was for sale three kilometres from their farm. They expected that they would have to pay at auction a price near the $8,000/ha that recent sales of good dairy land with good facilities and water rights attached were bringing. Competition was expected to be strong in the district.

Using the farmer-defined expected gross margin per cow and other costs, the preliminary analysis suggested that the new land would generate an extra total gross margin of $150,000 (250 × $600), plus $60,000 more from the home farm (that is, an extra 100 cows × $600), giving a total extra gross margin of $210,000. The expansion of the herd on the new and existing farms would involve extra permanent labour. The total extra permanent labour and management allowance would cost $100,000, with extra overheads of $20,000, leaving potential extra profit and debt servicing funds of $70,000 after tax.

The total extra investment would be $1.1m (that is, 80 hectares × $8,000 for land + $50,000 for plant + 350 × $800 for cows + $50,000 for infrastructure + transfer charges of $60,000). With $300,000 of equity capital, there would be $800,000 of extra medium-term debt (15 years at 8% per cent nominal interest). Around $70,000 per year would be required to service it. Collateral for the loan was to be based on 60% of the value of the new farm and on the 85% equity in the current farm assets.

The results of the discounted cash flow analyses were that with $300,000 equity capital, and paying $8,000/ha, the expected nominal rate of return after tax of this investment was 9.72%. Rate of return with full equity would be expected to be 8.3%. The accumulated net cash flow by year 15 was expected to be $1.5m. This represented a prospective build-up of net worth over the planning period of over $400,000 in nominal terms. If gross margin per cow was to be $500, not $600, over the life of the investment, nominal return after tax would fall to 4%, and financing arrangements would have to be changed.

These farmers believed that dairying had a good future, and as far as the farmers were concerned, the asking market price was not a barrier to their expansion plans. Their main focus was on growth and how they could finance it. They had few doubts about the economic merit of the investment and the potential of the planned investment to help them achieve their medium-term goals of growth in their net worth. They reasoned that if the investment could be financed, 'the economic side of things would look after itself'.

The prospect of a build-up in net worth over the medium term was the main objective in this land purchase decision. The economic soundness of the investment was a firm judgment the family had formed, based not on detailed analysis but on their overall knowledge of, and optimism about, the prospects of 'the industry', and on the results they achieved in their existing business. The family considered that as long as the investment could be financed, and building net worth was a likely result, this is what they would do because it was what they wanted to do in their business/farming career. They bought the farm.

## Mixed cropping–grazing property purchase

This land purchase case study was in a mixed farming region. In this area, land was considered to be 'tightly held', and there was considerable competitive pressure from part-time farmers. This farm family was facing a difficult situation in that while the current operation had been adequate for the parents to make a living and provide for their large family, one of their sons had returned from studying at an agricultural college and was determined to have a career in farming. He was working part-time at home and part-time on a nearby farm.

The original farm of 320 hectares had been expanded about 10 years before by leasing a further 120 hectares nearby on a long-term lease. The current lease charge was $70/ha. In good years, the 540 hectares of mixed cropping and grazing was capable of generating a total gross margin of $80,000, which left around $20,000 operating profit after the cash overheads ($15,000), and the non-cash overheads of depreciation ($15,000) and allowance for operator's labour and management ($30,000).

The business had some debt, but also substantial assets invested off the farm. The son earned $15,000 before tax per annum by working half-time off the farm. Also, the father was at an age where he would prefer to do less of the farm work and, with off-farm income, required only a small income from the farm.

A neighbouring block of land of 160 hectares had come on the market to be sold at auction, if not before. The seller was an elderly neighbour. The case study farmer had always had an informal agreement with this neighbour that if he did sell eventually, they would have the first chance to buy. The farmers knew that this land could be farmed using the same cropping–livestock system as the current farming operation. The existing cropping plant could be worked harder to handle another 160 hectares, with a small increase in annual repairs and maintenance – a cost included in the extra activity gross margin – at least for the first few years.

In this case, the farmers and the researcher worked through the preliminary farm management analysis together as part of the process of deciding on an offer price. The preliminary analysis went as follows: the annual gross margin per rotation hectare from this land was expected to be $200/ha ($75/acre), giving a total gross margin of $32,000. As few extra overheads would be involved, apart from shire rates on the land and some depreciation on fixtures, the net return from the land was expected to be around $25,000 after tax. This extra return would 'top up' the current part-wage paid to the son by the farm business. As well, the father was planning to do less work, and so would reduce his labour allowance accordingly.

Thus the whole business would end up with an expected total gross margin of $110,000 and an operating profit of $30,000 after cash overheads ($20,000), depreciation ($15,000), and with one full-time labour unit ($35,000) and a part-time labour unit costing ($10,000).

The owner was asking $2,400 per hectare. At this price, total capital required would be $400,000 for land plus 500 DSE at $20/DSE, $10,000, a total of $410,000. An operating profit after tax of $25,000 on 160 hectares, plus stock, represented an expected nominal return on capital after tax of 6% (that is, $25,000/$410,000).

The owners regarded an expected real return on marginal capital of 6% after tax as nearly acceptable considering the risk involved. Even though real capital gains were a distinct possibility in this area, possibly of the order of 2–4% p.a. real from non-commercial farming demand, the farmers said: 'Land values can go down, too, so we don't count on real capital gains. It would depend on when the land was sold, anyway!'

Still, the strong possibility of real capital gains helped the farmers to decide that they would be prepared to make a strong bid to buy the land. They decided to offer a price per hectare that would give them an expected real return per hectare of 7% after tax (before any capital gains). They offered $2,250 per hectare ($360,000). The $2,250/ha asking price was reasonably comparable with sales of similar land in the district in the past, though apparently there had been a couple of unusually high-priced sales recently that had raised the hopes of the seller. Though the initial offer was rejected, after some negotiation an agreement was reached to buy the land for $2,375/ha ($380,000 total). This represented an investment with an expected nominal return on capital of 7.92% p.a. after tax, with borrowings of $180,000 and $200,000 equity invested.

Financial aspects of this deal were straightforward. The vendor was willing to sell on vendor's terms of $200,000 deposit and the balance of $180,000 over five years in equal instalments at 7% p.a. nominal interest (slightly below market rates). With market rates of interest at 8%, this meant the buyers were getting the land for nearer to the $2,250 per hectare they initially offered. The annual debt servicing commitment was $13,000 interest and $36,000 principal. The deposit was financed from liquidation of some off-farm assets. The debt would be serviced out of the earnings of the off-farm assets and farm operating profit. The $36,000 principal repayments were to be met from assets and from liquidation of off-farm assets as required.

The DCF analysis indicates a nominal return after tax of 7.92% p.a. on marginal capital after tax, at $200,000 equity capital invested. At full equity, the nominal return after tax was 7.63% p.a. Annual real capital gains of 3% would raise nominal return on capital after tax to around 10% p.a.

The motivation of these farmers was to establish the situation where the father could work less and the son could become a full-time farmer. Financing arrangements of the purchase were on a sound basis, with some 'fall back' options if things didn't turn out as expected. For these farmers, return on capital was also of some interest in their decision. Further, they were well aware of the sound possibility of their land earning real capital gains, because the farm was located near a major inter-city highway and was in an area that was becoming increasingly attractive to urban and hobby farmers.

### *Conclusions from the land purchase case studies*

The main message from the case studies is that there are many aspects of every farmland purchase, and not all are amenable to inclusion in the economic models that are used to analyse the investment decision.

In all cases, the prices paid for similar land in recent sales played a large part in the farmers' thinking. This raises the question about the processes that were involved in forming the recent offer and acceptance prices of these sales of similar land. Presumably, the price of similar land played a role in the decision to accept the offer, too. Predominantly, all the case study farmers had clear views on the physical production possibilities of the extra farmland in which they were interested. They translated the expected extra production to some approximate ideas about the corresponding extra 'profit', usually extra net cash flows, that might eventuate. Profit was less important than cash, and they related the likely extra net cash flow to financial arrangements.

In essence, the farmers all knew well the results that they had achieved in the past, and expected to be able to farm with similar success on the farmland they were interested in buying. The implied expected real returns per annum after tax on extra capital featured in all the cases, to varying degrees, and ranged from 5% to 15% p.a. This was higher in each case than the return on total capital of the existing operation. Return on total capital of the existing operation was not in any of the cases a measure that was accorded as much importance as financing considerations. Expected return on extra capital was one indicator of the general order of magnitude of possible returns that could be related to other investments. The farmers knew that their return on capital would vary considerably around the estimates.

Family situation and medium-term goals seemed to be critically important in the decision to try and buy more farmland. Explicit consideration of inflation appeared to be absent. Expectations about inflation were included implicitly, though. The farmers used similar recent sales as a base for forming an offer price, as well as using current and recent past nominal commodity prices, production costs and interest rates in their thinking. Further, there was the reassurance of being likely to obtain some nominal capital gains. In every case, explicit recognition of cost of capital tended to be focused on nominal borrowing costs.

The case study findings accord with the findings of King and Sinden (1994) on buyers' bid prices. They found the following to be important considerations:

- children wishing to go into farming;
- the state of the farm and its productive potential;
- proximity to major population centres; and
- farm size.

The farmland investments analysed in the case studies were highly heterogeneous in their characteristics and in the expected net benefits to prospective buyers. Each of the case studies revealed significant layers of complexity associated with the valuation placed on the farmland in question. Each case seemed to have its own unique 'angle' on the land purchase decision. There were family situations, long-standing informal arrangements between buyers and sellers, tacit agreements between potential buyers, financial and administrative complexities, vendor finance at below market interest rates, risk considerations, marginal returns, development potential, government policy, compensation subsidies, growth objectives, beliefs, optimism and pessimism.

The market value method was the most common approach used. This method requires the potential buyer to explore the market prices received in recent sales of similar types of properties. This figure then has to be adjusted considerably for a number of important factors. These include:

- The method and conditions of the sale. When a seller has provided vendor finance at below market rates, the sale price has to be adjusted to an equivalent market price without vendor finance, or to an equivalent market price under the terms of the proposed sale of the land which is under consideration for purchase.
- The differences in the size of the property, the type and state of improvements (especially house), the management history (well-farmed and maintained, or run-down and neglected), and the different types and state of the soils, pasture and topography.
- The different timing of the past sales compared to the present.
- Changes in expectations about future commodity prices, inflation, interest rates, and the related local and international economic conditions.

Such adjustments are critical but are difficult. Also, only genuine sales are worth looking at. All up, the adjusted market value gives some indication of what a potential buyer might have to pay for a property. Like the discounted cash flow investment appraisal method, it provides further information to use in the decision.

Having generated information about 'should pay', 'could pay' and market prices of similar land, the potential land buyer is in a position not only to define prices of land that would be satisfactory, but also prices that should *not* be paid. The implied profitability of land in the future can be broken down into implied yields and prices that will deliver outcomes. If these are unrealistically high, then the land price is too high and the return required from the investment is unlikely to be attainable.

### *Adjusting land price for financing by vendor*

Farmland vendors sometimes provide finance on favourable terms to buyers. Favourable finance is an effective interest rate less than market rates. If the land sale transaction involves favourable financing, the sale price has two components:

- value of the property; and
- value of the favourable finance.

If sale terms include the vendor providing finance at an interest rate less than the market rate, then the sale price overstates the true amount paid for, and the value of, the property. The true market value of the property is the amount paid on reaching the sale agreement, plus the present value of the annual principal and interest payments to the vendor over the number of years required to repay vendor finance, estimated using market, not vendor-concessional, rates of interest.

## LEASING LAND

The key to leasing and sharing arrangements is the notion of each party being rewarded fairly in terms of their respective contributions of factors of production and their relative share of the burden of risk. With leasing of land, usually the landowner provides land and fixed improvements. With share-farming, the landowner will put in land and fixed improvements, and in some cases a variety of other inputs as well. Sometimes an agreement will involve a mix of cash lease and sharing some of the costs and income as well. For tax purposes the landowner in a cash lease agreement doesn't qualify as a primary producer, whereas the landowner in a share-farming arrangement is considered to be a primary producer. A lease of share-farming agreement has to set out how production and income will be shared, what resources and expenses will be put in, what improvements are to be undertaken, and any special conditions relating to such things as annual fertiliser applications, movement of livestock on to the property, restrictions on crop sequences, and repair and maintenance of improvements. With a cash lease the lessor may take none of the risk and the lessee may take all of the risk. With share-farming the risk is shared in proportion to the share of costs contributed.

There is much tradition and custom involved in the values that are decided in leasing and share-farming agreements. In different areas will be found different standard 'rules of thumb' or generalised starting points for negotiation, with adjustment made for the circumstances of each case. For example, share-farming agreements will reflect the different expected productivity of different pieces of land. Piece of land *A* might be expected to yield $400 gross return per hectare, while land *B* might be expected to produce gross returns of $300 per hectare. Share-farming costs in both cases might be $200 per hectare. The share-farmer of land *A* would need to receive half of the crop to cover expenses, while the share-farmer of land *B* would need to receive two-thirds of the crop to cover expenses. Further, the time when the money is paid for a lease will affect how much is paid. For instance, $100 after harvest in January would be equivalent to paying $92.50 in the previous June at 15% p.a. interest.

There are a variety of methods of calculating how much to offer to lease some land. One method is based on the minimum the landowner can accept and the maximum the tenant could pay. This requires calculation of the opportunity cost of the landowner's capital and actual costs the landowner is going to bear by leasing the land, and the expected residual net income from production after all the non-land expenses have been met. That is the amount of money expected to be left after all the operating costs, depreciation of capital, and the labour and management costs of the tenant, and the opportunity cost of any capital inputs of the tenant. The residual is the maximum before consideration of riskiness that the tenant could pay for the lease if the expected outcome was to eventuate. This method gives the maximum lease payment the tenant could pay and still expect to receive a market return on all the inputs to production.

Knowing the maximum the tenant could pay and the minimum the landowner could accept provides a basis for the necessary negotiation on a lease fee acceptable to both parties.

Inflation is always a factor to take into consideration, especially for medium- and longer-term leases. Some provision is needed for periodic adjustment in line with inflation. In the following example, the calculations are done for a one-year lease for the forthcoming year. The values used are the expected nominal values that will apply during the year. (That is, expected inflation is accounted for.) Note that as the landowner expects the land to increase in value in line with the inflation – that is, the landowner's capital 'earns' at the inflation rate. Therefore, the opportunity interest required is only the real component of a market interest rate.

Expected inflation = 3% p.a.
Real opportunity interest on land value @ $1,000/ha @ 4% = $40
Depreciation on improvements/ha = $5
Repairs and maintenance costs/ha = $3
Property rates and taxes/ha = $10
Insurances/ha = $2
Minimum risk lease payment to land owner/ha = $60

Note: The interest on land value is based on the market value of the land and the expected opportunity rate of return after tax, plus nominal capital gains from inflation of 3% p.a. in this case. The equivalent money in the bank might earn a nominal 7% after tax with no capital gains or inflationary gains on the asset. So, 4% p.a. after-tax and capital gain return on land with 3% p.a. inflation is equivalent to a total of 7% p.a. after-tax return in the bank.

The situation of the tenant is as follows:

Expected gross income from the 500 hectares of land = $200,000

*Less*

Operating costs = $90,000

Depreciation of tenant's capital included in the deal = $10,000

Income tax = $10,000

Tenant's reward for labour and management = $25,000

Opportunity interest cost of tenant's capital at 7% nominal = $5,000

Total costs = $140,000

Expected surplus potentially available to both tenant and owner (maximum tenant could offer) = \$200,000 – \$140,000 = \$60,000

Per hectare = \$60,000/500 = \$120/ha

Between the low-risk maximum the tenant could offer of \$120/ha and the low-risk minimum the landowner could accept of \$60/ha, agreement will be reached, depending on how the risk is being shared. In this case, the tenant is taking nearly all of the risk and so would expect to pay a lease price that is much closer to the landowner's low-risk \$60/ha than to the tenant's low-risk \$120/ha.

With leasing and sharing arrangements it is expected that higher risk will be associated with higher expected returns, and lower risk with lower returns. From the tenant's viewpoint, sharing has less risk associated with an undesirable outcome than leasing does. The amount of 'rent' paid with sharing varies as yields and prices vary, whereas with leasing, the amount paid is constant regardless of the yield and price outcomes. If prices fall, then the share-farmer gets a higher return from sharing than from leasing because the cost of the share 'paid' to the landowner also declines. With leasing, the tenant pays the full lease even though price received falls below the expected price; thus the share remaining and going to the tenant is less than with the share arrangement.

From the landowner's viewpoint, with share farming, the landowner may retain more control over the operation of the farm than is the case with the cash lease. Also with sharing, as the landowner is taking a share of the risk, on average returns should be higher. If prices rise, with sharing the landowner gets some of the benefit, whereas with leasing the landowner can miss out. The lease agreement tends to lag behind an improved price or yield outlook.

From the tenant's viewpoint the cash lease might mean that over time the rent is lower than with sharing because the tenant is taking all of the risk. Also, the tenant has full control (subject to the conditions of the agreement) and reaps all the returns on his or her own managerial skills. When prices rise, the tenant gets the benefit – in the short term, at least.

From the landlord's viewpoint, leasing has the advantage of providing a risk-free income for the coming year, and he or she doesn't need to know much about farming, be involved or take decisions. Also, less capital input is required, compared with most sharing arrangements.

Length of lease is always a tricky issue. Ideally what is required is 'short lease, but long tenure'. This means appropriate flexibility is needed to meet the landowner's need to be able to finish an unsatisfactory arrangement and yet provide the lessee with the incentives necessary to farm with the long-term productivity of the farm's resources in mind. Short-term leases require strict conditions to be made about the addition of any improvements or fertiliser and maintenance of existing improvements. Deciding on compensation for the carryover effects of fertiliser can be a problem. Sometimes for this reason the landowner pays for the annual fertiliser and gets a proportionate share.

Leasing non-land assets can be an attractive option. Leasing can be a way of reducing the amount of own or borrowed capital required when acquiring the services of an asset. A leased asset can be shown as an asset in the balance sheet, and the present value of the lease's obligations can be shown as a debt. This captures the effect of the leased asset on the liquidity of the business.

In the next chapter, ways of managing risk and uncertainty in agriculture are investigated.

## CHAPTER SUMMARY

*In order to analyse an investment opportunity soundly, economic, financial and rist analysis is needed. It is necessary to allow for the fact that resources have alternative uses (the opportunity cost), and benefits and costs occurring at different times in the future are valued differently because people prefer a dollar of benefits received now to a dollar received later.*

## REVIEW QUESTIONS

1. What does it mean to say, 'This project has a net present value of $1 million at a nominal discount rate of 10%'?
2. Establishing valid comparisons of the situation that is likely to prevail with an investment, and without the investment, is the key to judging the merit of an investment to change a farm system. Explain.
3. Benefits or costs received at different times have different values; thus they have to be adjusted to their equivalent present values. The adjustment process is called discounting. What is the basis for deciding on the size of the adjustment factor – called the discount rate?
4. Explain the land value criteria of what you *should* pay in order to obtain a defined expected return on capital; what you *could* pay according to ability to finance the investment; market values of similar property; and what you *might be prepared* to pay, all things considered.

5. A starting point in determining how much to offer to lease some land is to estimate the extra net return that can be expected from using the land after all the other resources have been rewarded at their market values. Consider the risks. The maximum that could be offered is defined. Then, think about the minimum the owner of the land to be leased would accept. This usually is the market opportunity cost of the capital involved, plus costs borne by the owner. Calculate how much to offer to lease a block of agricultural land that you know about.
6. Measuring what you can, and thinking hard about other important but hard-to-measure factors, is the approach to use in farm benefit-cost analysis. For a partial budget analysis of a change to a farm system, work out expected return on extra capital, and list other factors that warrant consideration.
7. How would you incorporate consideration of risk in a farm management analysis?

## FURTHER READING

Boehlje, M.D. and Eidman, V.R. 1983, *Farm Management*, John Wiley & Sons, New York.

Heady, E.O. 1952, *Economics of Agricultural Production and Resource Use*, Prentice Hall, New York.

Hopkin, J.A., Barry, P.J. and Baker, C.B. 1999, *Financial Management in Agriculture*, The Interstate Printers and Publishers Inc., Danville, Illinois.

Madden, B.J. and Malcolm, B. 1996, 'Deciding on the worth of agricultural land', *Review of Marketing and Agricultural Economics*, vol. 64, pp. 152–6.

5

# Managing risk and uncertainty

## WHOLE FARM RISK MANAGEMENT

Risk is the source of above-average profits – and losses. Farmers face many different types of risk, and most farmers manage risk adequately enough to stay in business for quite a long time. 'Risk' is a term given various meanings, but they always relate to the volatility of potential outcomes. In business management the core of the management problem is dealing with uncertainty. In decision making, how do we cope with knowing that we don't know what is going to happen in the future? One consequence of uncertainty that makes some decision analyses problematical is that the decision maker's goals are modified in response to the existence of this uncertainty. Included in the decision process about alternative uses of resources associated with differing degrees of uncertainty and risk is the reality that the goals themselves are modified by the existence of uncertainty. The nature and extent of this modification is determined by the decision maker's perception of where the decision lies on the continuum from risk (probabilities can be estimated and risk analysed) to uncertainty (no probability estimates are possible, uncertainty isn't able to be analysed), and by their attitude to these circumstances.

Risk is conventionally classified into two types: business risk and financial risk. Business risk is the risk any business faces regardless of how it is financed. It comes from production and price risk, uncertainty and variability. 'Business risk' refers to variable yields of crops, reproduction rates, disease outbreaks, climatic variability, unexpected changes in markets and prices, changes in government policies and laws, fluctuations in inflation and interest rates, and personal mishaps. These risks exist regardless of whether the business is operated on the basis of using only the owner-operator's capital or whether some borrowed capital is also used. The importance of business risk is in the possibility of planned-for yields, prices and costs not occurring and reducing the ability of the business to pay for the inputs used, to service debt, and to appropriately reward labour, management and capital. Financial risk derives from the proportion of other people's money that is used in the business, relative to the proportion of owner-operator's capital – that is, the proportion of debt (*D*) to equity (*E*), called gearing (*D*/*E*). The higher the ratio of debt to equity, the higher the gearing ratio, the higher is the financial risk faced by the business because of the operation of the principle of increasing risk. Thus, financial risk exacerbates business risk.

The important business-related risks and uncertainties are those that have potential for causing great good or harm, such as having a major impact on the owner's goals that matter most, including wealth and business survival. Strategic decisions that play themselves out over a run of years are the most critical in achieving goals such as wealth, business survival, consumption and leisure. It is sometimes argued that the medium- and long-term outcomes are merely the coalescence of numerous day-to-day and tactical decisions. While the effects of day-to-day and tactical decisions add to whatever cumulative outcomes eventuate, the strategic periodic big decisions affecting intensification, extensification, specialisation, diversification, enterprise type, gearing, land and machinery acquisition are the main determinants of ultimate wealth and business survival. Yet much risk research in agricultural economics has focused on the short-term risks of farming. It is often a run of related risk events that has the biggest impact.

From a practical point of view it is important, therefore, to monitor relevant changes in both domains: changes impacting on production efficiency and variability; and changes occurring in the wider, longer-term environment that might indicate a need to consider strategic change. A failure to watch shifting fortunes at a strategic level can allow a farmer to slide into a situation of focusing effort on the more and more efficient production of the wrong output mix for the farm.

## Objectives, risk and uncertainty

When key determinants of farm performance, such as yield and prices, are perceived to be towards the uncertainty end of the risk–uncertainty continuum, the manager has to accommodate the fact that there are no known ways of dealing directly with the uncertainty. When we face risk, we can hedge against it in various ways because it has dimensions that are believed to be understood. By definition, that isn't the case with uncertainty. The presence of uncertainty and the inability to hedge against it means that farmers in this situation are compulsorily speculating. Farming in Australia has been described as 'a particularly precarious form of freelancing' (Goldsworthy 2002, p. 200).

So, if a farmer knows enough to know that they don't know the probabilities of yield or prices over a defined planning period, what is the rational response? What should you do if you are being forced to gamble against unknowable odds?

Like many less-than-ideal features of farming, uncertainty characterises the environments of firms in other sectors, too. In the 'mainstream' or 'non-farm' strategic management literature the argument is put that, in essence, understanding the game enables the specification of meaningful objectives, as well as valid strategies. Broadacre farmers in Australia understand that it is an act of hope over experience to specify anything other than a medium-term range of required returns on investment, or similar performance objectives, due to the shorter-term volatility which their businesses face.

At the same time, it is an incomplete approach to simply recognise the volatility, do what you can and hope for the best. This approach to strategy isn't capable of informing decision making sufficiently. It is an approach that provides no guidance or constraints on choices of new enterprises to move into, or major changes to output mix to adopt.

Instead, the rational thing to do is to recognise that financial performance cannot be under tight, close control, due to uncertainty, and to accept that: (a) there will be considerable variability in financial performance; (b) long-run survival is the core objective; and (c) the focus of strategic attention should be the refinement of the farm's productive potential. This will ensure that the financial performance of the farm will be as high as possible persistently, and that the farm is well positioned to take advantage of opportunities that arise from time to time.

This approach involves identifying enterprises that are 'physically rational' for the farm and farmer, as well as 'financially rational'. Financially rational enterprises will be that set that seems to offer a level of volatility and range of financial outcomes over time that are most consistent with the farmer's attitude to risk. Included in the notion of financially rational enterprises are alternative steps a farmer may use, effectively as extensions to the enterprises, such as insurance or futures, to modify the effects of volatility.

The important point to be understood here is that uncertainty is common in farming and that its existence emphasises the fact that short-term changes in relative prices usually imply little for the farm plan. Strategy only has meaning when the appropriate long-term activity and farm plan is different from the sequence of activity indicated by short-term incentives. With uncertainty surrounding yield and prices, the farmer needs to deal with financial volatility with a long-term response. This will involve placing most farm management emphasis on controlling what can be controlled: farm physical productivity and choices about investments (that is, investment portfolio).

The uncertainty, and consequent variability in financial performance, facing farms means it isn't sensible to be highly sensitive to short-term changes in the operating environment. It is necessary to watch for changes that are medium-term trends that might occur in relative prices or production risk. Successful farm management involves a multi-period commitment to activities that, in the medium term, are expected to deliver desired financial outcomes. This approach is captured in farmers' oft-expressed disdain for 'chasing the market'.

The omnipresence of uncertainty means that farmers must grasp well the magnitude of what uncertainty implies for farm performance, and must consider ways in which matters that are under farmers' control can be modified to align their system with its associated risks and uncertainties with farmers' preferences for risk. Otherwise, the danger is that unwitting optimism may destroy the farm business. In farming, excessive optimism about the levels and stability of net returns involves an intellectual flaw.

## ANALYSING RISK

Appreciation of risk and uncertainty and its management is aided in all manner of ways by more information and by greater clarity of communication about the risk and uncertainty. The framing of questions is the key to answering them, especially in gaining understanding of risk and uncertainty. Gigerenzer (2002) talks of the 'miscommunication of risk – the failure to communicate risk in an understandable way' (p. 33). He makes a compelling case for presenting risk information, in terms of the arithmetic of events and populations of interest, using natural frequencies. This means, instead of saying 'This has a probability of 0.1', we say 'This could happen 10 times in 100 (or one time in 10)'. Such a simple step, applied to common risk decision situations, can do much to better the understanding of risk situations and so lead to more informed analysis and better decisions. The simple arithmetic of risk can bring clear perspective to probabilistic events.

Investors usually prefer more certainty to less, and less risk to more risk, while higher returns are always regarded as being more attractive than lower returns. Generally, investors will trade some risk for some return, though it needs to be remembered that high returns cannot be achieved without a corresponding level of risk.

It is useful to understand probabilities when making decisions on proposals that are risky. Probabilities are strengths of belief about an event happening, and are expressed as a rating from zero (no chance) to one (certainty). Probabilities can be stated as the odds of an event occurring. An event might be thought to have a probability of occurring of 0.4. This means that it is thought that there are four chances in 10 of it happening, and six chances in 10 of it not happening. In terms of odds, this is a 6/4 against chance. If it were thought to have 0.6 chance of happening and 0.4 chance of not happening, then this would be a 6/4 on chance (written as 4/6 against) – that is, it is less than 1/1, an even money or 50:50 bet. Something that was regarded as fairly certain – say, about 0.8 probability – would have eight chances of happening and two chances of not happening. This is a 4/1 on (1/4) chance, about as sure a thing as you could get. Odds of 100/1 against means that the person who is offering the odds believes that there is one chance of the event happening and 100 chances of it not happening.

The basic rule is that the probabilities assigned to two or more related events must add up to 1.0 (100%). When a coin is thrown into the air, the probability of it falling heads upwards is one out of two – that is, 50% or 0.5. The likelihood of it falling tails upwards is also one out of two, or 0.5. It is usual to expect an event with a 50:50 chance of occurring to happen half the time. That is, if someone had a fair coin and were to toss it many times, it could be expected that half heads and half tails would show up. Suppose that after a number of tosses there have been more heads than tails. Then someone might back tails in the hope that the next toss would show a tail. They would be mixing up two things: first, the undeniable truth that ultimately 50% heads and 50% tails will happen; and second, the reality that as the percentage of each outcome approaches 50%, the absolute difference between the number of heads and tails can still be very large. This is because a large number of tosses would have to have been made. The bettor can still end up losing, even after a large number of tosses of the coin. In terms of the total number of losing bets the bettor could still have backed many more losers than winners. The percentage of each outcome that occurs will certainly be edging towards 50%, but the divergence from 50% of each outcome (say, 49.5%:50.5% after a large number of events) will be a small fraction of a large number. It can all add up to more losing bets than winning ones, and could still be a large number of losing bets. The 'gambler's fallacy' occurs because the 'law of averages' or 'the 50:50 correction factor' works only over a very, very large number of events: so much so that the probability for each single toss of the coin remains the same. Many of the events about which farmers have to assess probabilities are like this. For practical purposes, the probability of a good season following a bad season is the same as the chance of a good season following a good season, or a bad season following a bad season. The coin doesn't remember.

A technique for analysing risky decisions is to calculate expected values of possible outcomes. For example, if there are 1,000 tickets in a lottery and the prize is worth $1,000, then each ticket has an expected value of ($1,000) by (0.001) chance of winning. The expected value of a ticket is $1. In this case, there are 1,000 possible winning tickets, each with a one in 1,000 chance of winning, each with an expected value of $1. The expected value of buying all the tickets is $1,000. Expected value is worked out as the sum of a series of possible outcomes such as 'good', 'most likely' or 'poor' seasons or prices. These are then multiplied by the probability of them happening, as estimated by the decision maker. So, instead of using a single value for yield of, say, 2 tonnes per hectare and price $200/t, probabilities can be used to work out an expected value for yield, as shown in Table 5.1.

If the most likely figure were used, it would be 2 tonnes per hectare. If outcomes were what had happened in the past 10 years, then the expected value of yield over this period would be 1.74 tonnes/hectare and expected price would be $218/tonne. Expected value is more useful than historical average, because historical average relates to a unique past, never to be repeated exactly, while expected value relates to what is believed might happen in the new, different and unknown future. The expected value is a composite figure which captures the situation that would apply if all the chances came at once. Or, it is what you would get, on average, if you made the same investment a large number of times, and the probability distribution of outcomes remained stable. If this was to be true, then profit over time will be greatest if the alternatives having the highest expected money values (EMVs) are always chosen. There are a number of

**Table 5.1** Expected value

| Season type | Yield (t/ha) | Years in 10 | Probability | Expected value (tonnes/ha) (Probability × Yield) |
|---|---|---|---|---|
| Best | 3.5 | 1 | 0.1 | 0.35 |
| Good | 2.5 | 1 | 0.1 | 0.25 |
| Most likely | 2 | 4 | 0.4 | 0.8 |
| Poor | 1 | 3 | 0.3 | 0.3 |
| Worst | 0.4 | 1 | 0.1 | 0.04 |
| Expected value (tonnes/ha) | | | | 1.74 |

| Price level | Price ($/t) | Years in 10 | Probability | Expected value ($) (Probability × Yield) |
|---|---|---|---|---|
| Best | 400 | 1 | 0.1 | 40 |
| Good | 300 | 2 | 0.2 | 60 |
| Most likely | 200 | 4 | 0.4 | 80 |
| Poor | 140 | 2 | 0.2 | 28 |
| Worst | 100 | 1 | 0.1 | 10 |
| Expected value ($) | | | | 218 |

limits to the usefulness of probabilities and EMVs. These are associated with the relatively small number of the main 'events' (seasons, prices, yields, investment decisions) involved in a farmer's career, when the 'laws' of probabilities work best over a large number of 'events', and the consequences of some events are far more serious for the success or failure of the business than other events. The small sample of 'events' means that good or bad 'luck' in terms of important variables in farming life has a significant effect over success or failure.

In farming, much is unknown and uncertain and thus the quality of the information about probabilities is often unavoidably poor. While probabilities can be used to give a more realistic measure than single-value estimates, when using probabilities the 'probability of the probability' problem arises. That is, the assumed distribution of probabilities is treated as though it is itself a certainty. More problems arise because the assumed relationships between the uncertain variables of an activity are also uncertain. There are some computer techniques that can be used to simulate runs of different seasons and/or prices to estimate probability distributions of outcomes on farm profit, finances and growth.

## Sensitivity testing, breakdown analysis and scenario analysis

A useful practical approach in assessing risk is to test the sensitivity of results to changes in critical variables, using the 'what if' approach – saying, in effect, if you do 'this', and 'this' happens, then the ultimate outcome(s) will be like 'this'. A small number of scenarios are defined – for example, poor yields and prices, high interest rates, or drought in years 2 and 6 of the 10-year life of the project. The decision maker weighs up all the information in the light of their own judgment about how likely it is that the various important events and outcomes could happen. Estimating breakeven levels of key parameters is useful in decision analysis. 'Breakeven' is the level of a key parameter(s) that needs to prevail for the decision maker to be as well off after the change under investigation as before the change. Breakeven numbers give decision makers something 'concrete' to focus on – the chances of exceeding the breakeven level are then assessable. This is a particularly useful approach when, as is often the case, there is little information about the likelihood and level of performance of a key parameter involved in a change in the farm activity.

Sensitivity analysis measures how sensitive the results of an investment analysis are to a change in one of the variables. The variable-by-variable approach involves listing all variables that are important in the analysis, identifying a range of possible values for those variables, and then calculating results using each possible value of the variable while holding all other variables at their expected value. This indicates the significance of a component of an investment. Another common approach to sensitivity analysis is to prepare the optimistic, pessimistic and most likely estimates for each of the variables. These approaches are useful for relatively simple analyses; however, they assume that variables are independent. In reality, variables are often interdependent. When this is the case, alternative scenarios can be used for sensitivity analysis. Rather than using combinations of variables based on optimistic, pessimistic and expected values, different combinations of values of key variables are studied. These combinations are known as scenarios.

Sensitivity testing and breakdown and scenario analysis force the decision maker to recognise the key variables and their possible implications; it brings to the fore the interactions among variables. These analyses also have important limitations: it is often difficult to get important information about alternative scenarios for each important variable, and it can be difficult to develop meaningful scenarios. In any case, usually, only a small number of scenarios can be sensibly considered.

## Simulation

Another technique for analysing the implications of some of the risk and uncertainty associated with a decision is called simulation. This involves developing a budget about the proposal and, using a computer program, drawing on values for key variables from probability distributions. This is sensitivity analysis that considers many possible combinations of variables, rather than a few estimates or a few scenarios. To do this requires a great deal of information. In this method the values for variables are chosen based on the laws of probability. The computer is used to perform a large number of calculations that reflect many possible combinations of variables. The result is a probability distribution of the project's cash flows that can be used to estimate a probability distribution of returns to capital, NPV, and so on. The expected (probability weighted) NPV or the benefit-cost ratio can be calculated. The variance of the simulated distribution can also be determined as an indicator of the likely volatility of the returns. The complications of this approach are that there are usually many variables with many interrelationships involved, and the relationships (correlations) between probability distributions of key variables can be complicated or, more often, not known. Further, variables can assume a wide range of values, with a continuous rather than a discrete probability distribution; and the relationships among variables and the probability distribution will change over time. Even more difficult is the fact that results in a particular year often depend on the results of a previous year. Simulation is expensive and unnecessary for simple problems, but sometimes it is a useful way to investigate complex problems and is valuable in research. However, simulation produces information about the range of likely results along with the estimates of the expected value, and sometimes combinations of variables are revealed which produce unexpected outcomes.

## Decision trees

A structured approach for analysing uncertainty is the decision tree method. This involves a graphic representation of the different possible outcomes of a decision, and probability analysis to calculate the overall expected value of the project (see Figure 5.1). The decision tree is generally used for projects that have costs or benefits in several time periods and where sequential steps or decisions may be required. However, they can become very complicated for projects with long lives or those that have a large number of possible outcomes. Using a decision tree,

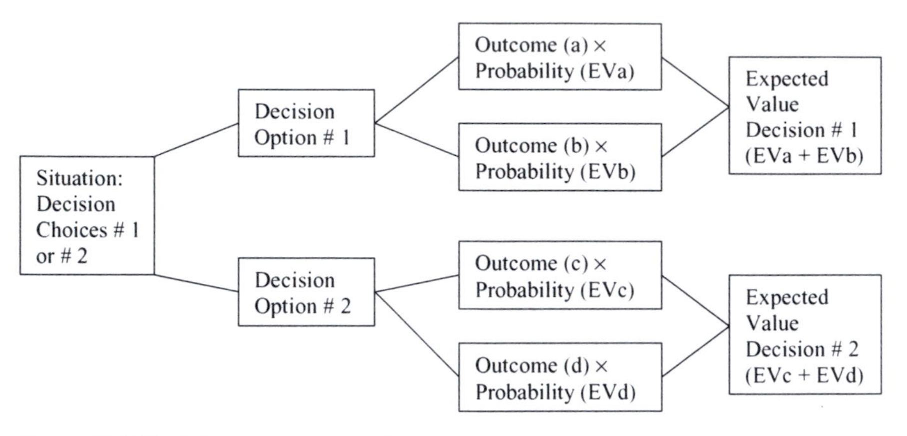


**Figure 5.1** Decision tree approach

different options for the project are analysed on a sequential basis. Possible outcomes in each period are identified and assigned probabilities. Outcomes and probabilities are shown on a flow chart that resembles tree branches, and results are then summed over time to calculate the expected value of each option for the project. The option with the highest expected value is the preferred choice. Again, the information needed to form useful estimates of probabilities to construct decision trees is often difficult to gather.

## MANAGING RISK

How can bad decisions about land purchase, farm development and machinery costs be avoided? Basically, simple farm management budgets can be used to identify what not to do. That is, prices of land can be identified which imply a run of unrealistic, over-optimistic, unsustainable, unfinancial and otherwise inappropriate levels of the key parameters of output prices, yields, inflation rates, capital gains, interest and discount rates over the planning period. Thus prices 'not to pay' can be identified, as also can prices of land that represent good economic returns based on realistic expectations, including riskiness. Similarly for machinery and development decisions, particular options can be identified as being 'not wise investments' nor 'good value'. If equity is low, it is sensible to use high incomes to lift equity and so (hopefully) weather the next storm. Given a current high equity, a number of options are available to the farmer. An extra 20–30% of land can be added to the existing farm when the pressure goes off the boom, or during a bust at the start of the inevitable recovery. Investment in proven technology to further intensify the production process is another approach.

Often, a decision has to be made about specialisation versus diversification in activity mix. A traditional view is that diversification is the key to coping with overall risk, by spreading the nature of the risks a farmer faces. There is sense in this, particularly in taking a portfolio view of investment and spreading sources of income more broadly rather than relying solely on agriculture for income. There is sense in diversifying operations spatially. There is sense in diversification of activities within a farm, provided the activity incomes are not in some way correlated. However, the scope for diversification is limited when activities are complementary. For example, some wool and mutton production might be a necessary part of a grain-growing operation, regardless of relatively less attractive returns from wool and mutton, compared with some other possibilities on the farm. Diversification is usually seen as a wise management strategy to reduce the problems of variability of yields, prices and income. Those areas of Australia that readily lend themselves to producing a wide range of commodities are often the most desirable areas to farm. However, many farmers are not in this situation, so their chances of diversifying are limited. Appropriate diversification on and off the farm can make up for fluctuations in commodity prices. Diversification can make income variability greater if the various activities are affected by similar seasonal and market forces. In this case, prices and yields move in the same direction at the same time.

Less often recognised is that specialisation (doing what you do very well/better), too, can be a form of risk management. Specialisation can result in greater efficiency of production and better quality and reliability of product, thus providing some 'insurance' against some of the price and climate risks a farmer may face. In an equally 'contrary' way, diversification can cause an increase in risk, if the diversification is out of a field the farmer knows and does well and into a field that he or she doesn't know or do well. One of the lessons from the financial disasters of the so-called business entrepreneurs of the 1980s is that diversifying out of what you know and do best into a new field can be the main cause of subsequent financial disaster.

## Coping with drought risk

There is a vast array of management actions aimed at combating the impact of the reduced cash flows, feed shortages and falls in value of assets that result from drought. The main on-farm strategies for grazing operations include the use of stored fodder, progressive sale of different classes of livestock, agistment of some stock, purchase of grain and fodder during the drought, letting stock fend for themselves, or a mixture of the above. Depending on prevailing and expected prices, equity, debt, type and liquidity position, the cost structure, and the expected likelihood of drought occurring, each of these strategies has a place and is employed. The problem, of course, is estimating when a dry period is going to turn into a drought, and if it does, how long it will last. Rainfall distribution, evaporation, farm activities and their needs all determine if a farm is in drought. Farm activities in any area evolve, over time, to fit the varying climatic conditions of the area. For animal enterprises, the breeding, mating, stocking, weaning and feeding management strategies evolve. For cropping, the types and varieties of crops grown, crop sequences, tillage practices, timing of operations and harvesting systems are all adapted to suit the climate. So, drought isn't simply a severe water shortage, but an abnormal, unexpected, severe water shortage.

To cope with drought risk, it is essential to make decisions early, then monitor things closely – planning, timing and execution are the keys. Surviving a drought depends as much on actions taken in previous years, as on actions taken during the drought. An important part of preparing for drought is to exploit good years with the aim of improving the financial ability of the farm business to withstand the certain future drought. As well as aiming to be financially 'secured' against drought, common sense dictates that cash surpluses from good years should be invested in adequate water supplies, and perhaps in fodder reserves. The ultimate effect of drought is to reduce equity or net worth via loss of income, cost of feed and loss of livestock. Droughts force farmers into making decisions that determine their future options, and even their livelihood. The obvious uncertainties of droughts force farmers into an awareness of forward planning. There can be no specific prescriptions for drought survival and recovery. Drought decisions will ultimately be based on an individual's situation and experience. To do nothing in anticipation of the inevitable drought can be very costly. To do everything may be equally so. The seriousness of the adverse effects of drought can be ameliorated by management with an emphasis on the 'long view'. Farming with a long view is a more realistic option for those businesses that are not too highly geared, and which have sufficient physical and financial resources to plan forward on the basis of expected average cash flow over a number of years, and whose survival isn't absolutely reliant on near future cash flows. Apart from protecting incomes, Australia's soil and livestock resources will be protected most efficiently if farmers undertake such long-term strategies.

## Pasture development and cropping risk

Pasture development decisions need to incorporate some allowance for the risk of establishment failure before the investment gets a pass, because the initial pasture establishment process is expensive and is so often fraught with danger. This is particularly so in the low winter rainfall (for autumn sowing) and short spring/unreliable summer rainfall regions (for spring sowing). The decision to spend several hundred dollars per hectare on fertiliser, sprays, seed and cultivation, and sowing, plus lost grazing, can be quite risky in such regions. Nevertheless, pasture development usually has to be undertaken to ensure long-term viability. A continuation of apparent progress in short- and medium-term weather prediction will help to improve the timing and success of pasture establishment processes. The same types of benefits apply in cropping, with some savings from reduced losses from total crop failures.

In cropping, the production – and risk – decisions about activity mix are often straight-forward because for any paddock (or crop area), it usually involves only a couple of technically sensible alternatives competing for essentially the same resources at the same time. The choice for a paddock or crop area for the current crop year can be based on expected gross margin from activities and expected total gross margin from the whole farm activity mix. The decision is made with an eye on the susceptibility to yield, price and cash flow failure, and on what a particular activity on 'this' area might mean for activities on the same area in the next year or two. Crop farmers have considerable scope to use forward pricing and futures contracts, and have various insurance possibilities. They also have the possibility of selling into a number of different markets and pools, and doing so over a range of times. Pooling of risk through cooperative behaviour by producers has long been a standard approach to risk management in what is one of the riskiest cropping environments in the world.

## Stocking risks

The grazing situation is more complex. Essentially, the key production and risk questions are what stocking rate, and what mix of animals, to choose. The risk-related questions in animal production are complex because of the intricate dynamics of the pasture–animal–animal product complex whereby the pasture production determines the stocking rate and animal product, and the stocking rate determines the pasture production. Also, the decisions are complicated by the relatively long production period involved. It is often overlooked that intensifying production – say, by increasing stocking rate – affects marginal output and variable and risk costs, and sometimes overheads as well.

Judgment needs to be made about the marginal production per hectare for each particular paddock, and about the new set of variable and risk costs that might be involved with this new, more intensive system of animal production. Intensification can increase both the absolute size and the variance of net income from an activity and the whole farm. To propose increases in stocking rate without corresponding increases in risk costs will lead to an incorrect decision. As stocking rate increases, risk costs increase, too. To help cope with price risks, graziers have the choice of using futures contracts in beef and wool (though these are used little), of direct forward contracts and relationships with lamb and cattle processors, and of using a range of selling methods, markets and marketing times.

## In sum

In practice, farmers take many different steps to place their business in a 'risk situation' that gives a good chance of long-term survival. These include (among other things):

- being good at the technology – that is, operationally efficient and timely;
- the business not being highly geared;
- keeping overheads low relative to output, gross income and total gross margin – achieved through keeping production up, judicious expansion, tight control of costs, prudent investment in machinery and stock, and not having too many family members trying to get a living off the place;
- specialising or diversifying (either one can be the best risk strategy, depending on the case at hand);
- putting good year surpluses into good investments. (Depending on the state of development of the farm, the investment with the highest return after tax on marginal capital could be on the farm or off the farm. Beyond some reasonable level of farm productivity, sound investment off the farm is prudent. However, it has to be in something that the investor understands.)

- being well aware that intensifying farm businesses will increase the average net income and the variability of net income over time; and
- taking a portfolio view of total investments.

## THE ROLE OF FUTURES MARKETS IN MANAGING EXPOSURE OF FARM BUSINESSES TO VOLATILE PRICES

Successful marketing requires, in general, gathering, interpreting and acting on information in an exploitative way. Using futures markets requires considerable business sophistication, flexibility and attention to detail. Australia has futures markets for the agricultural commodities wool, wheat, barley, sorghum, canola and beef, and Australian cotton growers trade on the US futures market. Major financial institutions also offer 'over the counter' products that enable farmers to hedge price risks in a number of commodities without actually undertaking the futures trading activity themselves.

A futures market is a market in which contracts (not the actual commodities) are bought and sold for the future delivery of commodities. A futures contract can be defined as 'a legal contract enforceable by the rules of the Futures Exchange, to deliver or accept delivery of a definite amount of commodity of specified grade, during a specified month at a specified price'.

The key to understanding the operations carried out using futures market instruments is the widely used definition of 'a hedge in futures is a temporary substitute for a later transaction in the cash market'. This is the essential starting point for understanding futures. So, what happens?

> It is January and a primary producer expects to have some commodity X to sell in six months' time, in June. In January, prices for commodity X are quoted as being $100 for the coming June on the futures market. The farmer is worried that the price for commodity X might fall, and would like to be sure to get $100 for his commodity X. So he sells a June futures contract – effectively, an agreement to deliver commodity X in June and to receive $100 for it. In June, prices for commodity X have fallen. The farmer takes his commodity X to the physical market and sells it for $50. On the same day, he goes to the futures market and buys back his futures contract for commodity X, which costs him only $50.

> (That is, instead of delivering his commodity X to the futures market, he cancels out his obligation to do so by buying an equivalent contract back.) The physical and futures prices for commodity X are the same at this time (both $50) because on delivery day the futures and physical market prices converge. This happens because commodity X could, if necessary, be delivered to meet the future contract requirements. The competitive arbitraging behaviour of buyers and sellers means that prices for the same product in two markets become approximately the same. The overall result is that the farmer has received $50 for his commodity X and made $50 profit on futures trading. He has avoided the effects of the price fall and in total received the $100 for his commodity X that he wanted to protect.

Futures markets are primarily concerned with the formation and discovery of prices through time. In a futures market, forward prices are determined for commodities based upon each individual's interpretation of the information available at any one time. A basic concern of futures is to provide a hedge against price risks. They are markets that enable risk to be shifted. Futures trading occurs in a central market, which in Australia is the Sydney Futures Exchange or the Australian Stock Exchange.

Futures contracts call for delivery of specific grades of the commodity at a specific location. However, contract rules generally allow the substitution of grades at predetermined discounts or premiums. In futures trading, there is a clearing house which keeps records of its members' dealings. The clearing house removes the individual responsibility of one member to another. Each member is responsible to the clearing house for their net position. Deliveries are seldom made because the responsibility is discharged before the delivery month by taking an offsetting position. A purchase cancels a previous sale and vice versa.

It is important to note that futures trading is in contracts, not in the physical commodity itself. A seller of, say, one wool contract can decide to deliver the goods, and will be paid at the contracted price. Similarly, a buyer of a contract can let the contract mature, and must then receive delivery of the goods and pay the price stipulated in the contract. But actual delivery of the commodity occurs in less than 2% of all contracts taken out. Most contracts are cancelled by offsetting transactions made prior to the time the contract matures. Usually, futures contracts are 'closed out' as soon as the physical transactions are completed. For example, a woolgrower may have hedged against a price fall and sold futures contracts for the expected wool clip. When the clip is sold, the futures contract is 'closed out' – contracts are bought back, thereby offsetting the contracts that were sold previously. The hedged price is protected by carrying out the physical and futures trades at the same time.

Futures markets are derivative – they derive from other markets. Futures and physical market prices move in a similar manner because they are influenced by the same economic factors, and because it is possible to actually deliver on the futures contract. Commodity, futures and physical prices converge as the delivery (spot) month is reached. This relationship between the physical and futures markets means that if a loss is made on the cash market, a profit can result in the futures market transaction. The net result is that some 'planned-for' profit can be achieved – approximately.

A successful futures market requires large numbers of buyers and sellers to provide liquidity and balance. No longer is the need for storage of seasonally produced commodities the only reason for futures markets working, as futures markets have emerged for other commodities, and many storable commodities function without future markets. In the past decade, many futures markets have developed in Australia, and worldwide, for non-commodity instruments such as interest rates, exchange rates, share price indices, and so on.

A successful futures market needs:

- contract terms and commission charges that attract sufficient use of the futures contract;
- handlers of the commodity, who must have reason to make substantial use of the futures contracts as temporary substitutes for contracts that they will make later – that is, for hedging purposes;
- the possibility of attracting speculation to provide a liquid market; and
- adequate public recognition of the economic usefulness of the futures market.

In sum, futures markets serve three basic functions. These functions are:

- as a major price-discovery mechanism;
- as a risk-shifting opportunity for producers and users of commodities with regard to adverse price movements (hedging); and
- as a means to speculate.

The price-discovery mechanism of the futures exchange is important for many people. These prices reflect all current information available in the marketplace. The concept of 'price discovery' is concerned with the process by which buyers and sellers arrive at specific prices, and not with the way economic forces determine the actual level of prices. In a futures market, the views of many buyers and sellers are focused on a single market. They have diverse objectives, but there is a continuous appraisal of the price-making forces.

A major contribution to efficient economic activity by futures markets is the increase in the quantity, timeliness and value of information, and thus in the accuracy of prices. There is always an incentive for traders to actively seek out more and better information. The success of speculation on the market is based primarily on expertise – that is, the ability to consider the decision-making process of others involved within the market. As such, speculators are in constant need of new information, information which thereby improves pricing. Futures markets also have a role in rationalising storage decisions. This can help to stabilise seasonal prices. Once all aspects of the operation of futures markets are considered, the conclusion is that even those participants in production and marketing, but not directly involved in the futures markets, can benefit from the operation of the futures markets.

The main features of the operation of futures markets are summarised below, with explanations given of the common trade jargon.

- Futures are markets in contracts, not physical commodities.
- Buying and selling futures contracts is one way of reducing uncertainty about a commodity price to be received at some future time. For example, a producer with a commodity to sell in the future can sell contracts to 'lock-in' a certain, acceptable price at some time in the future.
- Because contracts can be fulfilled by delivery of the contracted commodity (of specified quantity and quality), the cash and future market prices of the commodity converge as the futures contract expiry date approaches. If cash and futures prices are sufficiently out of line, then the futures contract could be settled by actual delivery of the goods. On the last day of trading for a contract, spot and futures prices will be approximately the same.
- Cash and futures prices tend to move together. This is because the cost of storage of the physical commodity over a period is a major element of the difference between spot and futures prices. Active buying and selling between

both cash and futures markets to profitably exploit any discrepancies around this cash–futures price relationship causes prices to be bid up or down, and means that the cash and futures prices tend to move together.

- A trader may establish a market position by either buying or selling. It isn't necessary to buy before selling. To be 'short', contracts are sold (obligation to deliver) and not covered by purchases – that is, more futures contracts are sold than products purchased. There is simply a contractual obligation to deliver the specified commodity at a specified price and to receive the agreed-upon price. The short position, however, can also be satisfied by buying back the contract as well as by making delivery. (This net position with the clearing house is zero.) To be 'long', a trader has purchased contracts (obligation to accept delivery) not covered by an equivalent amount of sales. He or she is obligated to accept delivery and to pay for the contracted position unless the long position is offset with appropriate sales.
- The futures transaction complements the cash transaction. Futures market prices in the delivery month are very close to the cash market prices. The holder of a sold contract sells the actual commodity on the cash market and settles the futures contract by buying back a sold contract. If the price received on the cash market is less than the budgeted price (a loss on the sale of the physical commodity), then the futures contract can be bought back for less than it was sold for (a profit on futures), and the budgeted return will approximately be achieved. Likewise, if it costs more to buy back the futures contract than it was previously sold for (a loss on futures), the physical product will be selling for more than was anticipated on the cash market, and again the net outcome will balance to approximate the budgeted return.
- If a futures contract price movement is unfavourable to a trader and exceeds the deposit (which all traders must lodge with the exchange), then the broker requests a margin from the trader – that is, a margin call. These margins are an assurance to cover the potential loss from closing out the particular futures market position. The outcome of a transaction in the futures market (which is offsetting – short position followed by purchase of product, or long position followed by a sale of product) depends on what happens to the price of the contract while it is held (ignoring brokerage costs). To cover the contingency of adverse price movements, traders are required to make a margin deposit to prevent default.

- Suppose the initial margin on a contract is $0.15 per unit of commodity ($750) and the maintenance margin is $0.10 per unit. As prices of a particular delivery month's contracts fluctuate, the equity (the amount the trader will receive when the contract is closed out) of the trader fluctuates. Now, if the trader is long (that is, has purchased futures but not yet arranged equivalent sales), then price increases are favourable. He or she will be able to sell the contract for more than the price at which it was purchased. This price movement would increase the trader's equity. On the other hand, if prices fall, the trader's equity is decreased. In this example, if price falls more than $0.05 per unit of commodity, the trader faces a loss on futures and the maintenance margin is breached. The trader is required to provide additional funds to restore the initial margin. Otherwise, the broker sells out the position. Conversely, short traders with net selling positions (who are worried about price rises) are required to increase their equity by further deposits as the price rises. As the price rises, they face having to buy back contracts for more than they sold them and suffering a loss on the futures trades. While traders may be confident about the eventual outcome of a contracted position in the market, their capital base needs to be such that margin calls due to adverse short-term price movements can be met.

## Key terms

**Arbitrage** Buying on one futures market and selling on another market elsewhere to take advantage of price differences between the two.

**Basis (or spread)** The difference between the cash (spot) price and the futures price.

**Bear** One who considers that the price of a certain commodity is about to fall.

**Bull** One who considers that the price of a given commodity is about to rise.

**Buying a futures contract** Taking out a futures contract to buy a commodity. More accurately, it is agreeing to buy a commodity at a later date.

**Contract** A futures contract is a legally enforceable agreement to buy or sell a specific grade of a commodity at a specified future time and specified price. Contract terms are highly specific as to both quantity and quality. Standard wool futures contracts are for 2,500 kilograms clean, or approximately 20 bales greasy, of 22 micron quality, of good to average length. There are associated premiums and discounts relating to variations from these characteristics.

**Deposits** A buyer or seller of a contract has to lodge a deposit for each contract entered into. Currently, for cattle futures the deposit is $600 per contract; for wool it is $400.

**Exchange** Provides trading and technical facilities for the operation of the futures markets. The main functions of the exchange are to enforce contracts, register transactions, act as a clearing house, and enable centralised trading of standardised contracts by open auction.

**Hedge** When a commodity holder contracts to buy or sell a commodity at a future time, at a specified price, in order to reduce the risk they face from an adverse movement in the cash price they may receive.

**Hedgers** Generally, traders involved in the production or marketing of the physical commodity. Hedgers are mainly concerned with protecting themselves against adverse price movements. They could be sellers of futures contracts (for example, primary producers) or buyers of futures contracts (for example, a wholesaler or retailer of goods, flour miller or grain merchant).

**Longs** Buying a contract is called a long, or going long.

**Loss on futures** If a seller buys back a contract at a higher price than the original was sold for, or if a buyer sells back a contract at a lower price than was paid for it, a loss on futures occurs.

**Margin** Payment made by a trader to the broker when a futures contract price moves unfavourably. A $0.01 per kilogram adverse price movement on one beef contract (10,000 kilograms) requires a $100 margin call to be met (at least once the trader's initial deposit has been exceeded).

**Profit on futures** If a seller buys back a contract at a lower price than the original was sold for, then a profit on futures is made. Similarly, if a buyer sells a contract at a higher price than the original was bought for, then a profit on futures is made.

**Selling a futures contract** Opening a futures contract to sell a commodity. More accurately, it is agreeing to sell a commodity at a later date.

**Settling a contract** This can be done by making or taking delivery of the commodity, but it isn't usually done. More commonly, a contract is settled by making an opposite, or offsetting, transaction. That is, a hedger who has sold a contract will close out by buying back a contract; a buyer will close out by selling back the bought contract.

**Shorts** Selling a contract is called a short, or going short.

**Speculator** Someone who accepts risk. They in effect assume the hedger's risk of price movements and are aiming to profit by the returns to bearing risk and/or by the returns from trading based on accurately forecasting future price movements and price levels.

**Spot price** The current cash price.

## Relationship between cash and futures prices

Understanding the relationships between cash prices (in the actual market) and futures prices for the delivery months specified in futures contracts is particularly useful. Grains are the best example of a seasonally produced, continuous by stored commodity. They have an annual harvest time and need to be stored between harvests so that they can be consumed throughout the year. The annual average price is determined by annual demand and supply conditions. A set of monthly prices is associated with the annual average. The lowest price for seasonally produced crops occurs at harvest, with prices rising throughout the year to cover the costs of storage. The difference between cash prices and futures prices may be positive or negative. This difference is called the basis. In defining the basis, the cash price is for the quality, location and delivery conditions specified in the futures contract. The basis narrows as the designated month of delivery approaches. This narrowing basis reflects the decreasing cost of storage as delivery time approaches. Call *PF* the futures price and *PC* the cash or spot price. There are two possibilities:

- $PF > PC$: *Positive basis*. This describes the situation where the current futures price is above the current cash price, as would be the case at harvest time with a commodity that is harvested seasonally.
- $PF < PC$: *Negative basis*. This describes a situation where the current cash price is above the futures price. The *PF* will be less than *PC* when current stocks are small relative to expected supplies. For example, at the end of a crop year, the price of futures for the harvest month for the new crop may be below the current cash price.

Merchants will hold stocks only if benefits are expected to equal or exceed the costs of storage between two points in time. The difference between the price for a future delivery month and the current cash price (or between the prices for two delivery months) defines the expected benefit from storage. That is, the basis may be considered as the price of storage.

This relationship holds because of the choice of accepting or making delivery of the commodity on the futures market by traders who have bought or sold futures contracts. The stockholding operations of traders in seasonally produced commodities are facilitated by futures markets. Coverage of the price of storage ($PF - PC$) can be assured through hedging.

> Suppose on 1 October the following price relationship existed:
>
> - cash wheat at $200/tonne; and
> - April wheat futures at $220/t.
>
> The potential return from storage is $20/t, and a grain trader sees this as sufficient for profitable storage. The trader buys wheat at $200/t and sells an equivalent quantity of April futures at $220/t. There is a return of $20/t to be had for storage. If the cash price is $220/t in April (as also will be the futures price then), then the trader sells the wheat he has in store (at a profit of $20) and buys back the futures contract for $220, at no gain or loss. If instead of being $220/t in April the price declined to $190, the trader still gains $20 because:
>
> - he sells wheat for $190 (loses $10 on cash trading); and
> - purchases futures contracts at $190 (gains $30 on futures trading).
>
> The $10 loss on the cash transaction is offset by a $30 gain on the futures market transaction. The direct costs of storage are constant whether prices move up or down. In addition, the trader gains the flexibility to sell the stored wheat privately that could enable him to do better than this.
>
> The typical trader has no intention of holding the entire stock through to April to sell on the physical market or to make delivery to the futures market at the then prevailing price. Rather, traders have an eye for any opportunities that may arise to sell wheat at a gain. This possibility arises because the dealer has grain of a quality or in a location of some special advantage to a particular buyer and the seller. The futures trade protects their position but doesn't dictate their actions.
>
> If the trader is faced with $PF < PC$, then the incentive isn't to store. An opportunity to make a contract for delivering grain in some future month might arise even though a trader might not have any grain in store. A forward cash sale can be accompanied by the purchase of futures

contracts to ensure that the purchase price of the grain is compatible with the forward sale price and a margin for the trader is protected: a buying hedge. For example, a trader might contract in March to deliver wool at $10/kilogram in November, and at the same time offset this forward contract by the purchase of December wool futures at $9/kg. If in October the merchant buys wool at $9.50/kg (to deliver on the forward contract), and sells December futures at $9.20/kg (to cover the hedge), the trader has gained $0.50/kg on the cash transaction and lost $0.20/kg on the futures market transaction for a gain of $0.30.

To summarise, both positive and negative differences between the futures price and the cash price provide incentives for particular types of transactions by merchants. The positive difference reflects current stocks being high and provides an incentive to store stocks. A negative difference between the futures and cash price reflects current shortages of stocks relative to future supplies and a disincentive to carry stocks, with an incentive to make forward contracts covered by a buying hedge.

The forces influencing price are continually being evaluated in a futures market. The level of prices is more variable than the variations of price between delivery months, which tend to move up and down together. Because new information tends to occur randomly, price changes appear to have elements of randomness.

A futures market links cash and futures prices through time. For example, consider information just available about an expected shortfall in production of a commodity such as wool or grain. This doesn't mean only a rise in the futures price of the delivery month. Both futures and cash prices will respond to this new information. All prices will be raised until the delivery month, because people will retain more wool or grain now in the knowledge that they have less available and will be buying less in the future.

## Hedging and speculation

Hedging means the hoped-for profit can be achieved, approximately. The price is protected by the futures contract, and losses from falling prices are minimised. However, if prices rise and the producer is hedged, speculative profits are forgone. This is why making a strict distinction between speculative and hedging transactions on futures markets isn't altogether accurate. All transactions involve taking some view about current and future prospects – they are part of the opportunities available to traders on markets. Most hedging doesn't have the objective of pure risk aversion or pure price insurance. The hedger knows that the basis tends to narrow as the delivery month approaches. A positive basis (futures price greater than cash price for different times) encourages selling hedges and favours holding stocks. A selling hedge ensures a minimum return from stockholding, but it removes the opportunity of earning larger returns when price rises above the hedged price. The buying hedge works best with inverse carrying charges (futures price less than cash price for different times) and provides protection against price increases. However, it precludes profits if price declines below the hedged price. The decision to hedge also involves implicit views about the course of prices in the market. Therefore, underlying the use of hedging are certain 'speculative' elements.

Speculators take market positions with the expectation of making a profit without taking offsetting positions in the cash market. They enter the market to buy risk off hedgers wishing to sell risk. Speculators contribute liquidity to the market in two ways: by absorbing seasonal hedges (typically short), and by their willingness to buy or sell at or near existing prices (as performed by scalpers). By performing these functions, speculators prevent individual transactions from greatly affecting prices. When short hedging is large, it has to be offset by long speculation. Long hedgers cannot be expected to enter the market to accommodate all the short hedging requirements at the right time and in the right quantities. Speculators play various different roles, according to differences in perspective and timing, but in all roles they provide the liquidity and willingness to buy risk that other participants want to sell. Speculative activity is critically important in responding to temporary imbalances among buyers and sellers in the market.

Futures markets have zero sum returns (gains equal losses), so that not all traders are getting rich on the profits from futures markets. The evidence suggests that a lot of players lose a little and a few players win a lot – that is, mostly the small, non-professional speculators are the losers. This would be expected, as they don't have sufficient market information on which to base their operations. It can be thought of as a form of gambling for them. There are other cases where the attraction of potentially large gains results in many small losses. In summary, speculators are trying to make profits, although many don't succeed. In the process, they add to market liquidity, which assists hedgers and aids in the process of price discovery.

### Influence of futures trading on cash prices

There is a tendency to think that trading futures contracts must somehow 'adversely' influence cash prices – say, by short futures trading increasing the frequency and magnitude of the variation in cash prices. A futures market is a price-discovery mechanism interpreting economic forces: it is these forces that determine the level of prices. And, like any other price-discovery mechanism, the performance of a futures market depends on a lack of market imperfections. It is more likely that futures markets stabilise prices because they facilitate the carrying of stocks and allow better production planning by providing relatively stable forward prices. Without futures trading in inventory-hedging markets, too little would be stored at harvest, with consequent lower prices, and then higher prices throughout the year.

Speculators, because they are attempting to capitalise on price changes in the market, can reduce the market's instability. By buying low and selling high, speculators can reduce price fluctuations. It is tempting to believe that speculators are able to drive prices up (hurting consumers) or down (hurting farmers). It is possible, in principle, for speculators to 'corner the market' by buying in the cash market and futures market, thereby squeezing those who have to make delivery and find they cannot buy the commodity to deliver on the contracts they have sold. This is illegal, and market reporting rules and the scale required tend to stop this activity.

The discussion so far has been based on commodities such as grains or wool, where the inventory-hedging function of futures markets is important. The size of the basis (the difference between spot and futures prices) can be interpreted as a price of storage. This role of futures markets has had to be reinterpreted in recent years because of the evolution of markets that don't have the same characteristics as seasonably produced, but continuously consumed, commodities that require storage within and between seasons. In these newer markets, inventories and inventory hedging play a lesser role. Futures prices in these commodity markets where storage is less significant are less closely tied to current prices for the same form of the commodity. This is because there won't be the same link over time through storage operations if commodities cannot be stored.

## OPTIONS

Farmers are in the position of wanting to avoid the effects of a fall in the price they receive for the products they sell, while at the same time not wanting to miss out on any increases in prices of their products. Options are a derivative of a futures market. Options are a way of taking out some 'insurance' against the price fall, while keeping the chance of cashing in on price rises. Note that options markets are not available on the Sydney Futures Exchange for wool or beef at present, mainly because there is an insufficient volume of trading on these futures markets. Trading options is possible for commodities such as wheat and cotton, which are traded on the US futures markets.

An option is an agreement between two people that gives one person the *right* to do something, without that person being *obligated* to do it. The option, or choice, is to buy whatever is specified in the option, or to sell whatever is specified in the option. An option that gives the right to buy something at a later date is called a call option. An option that gives the holder of the option the right to sell something at a specified price before a specified date is called a put option. Put options are of most interest to farmers intending to sell their production. In the market where options are bought and sold, the option buyer is called the option holder and the option seller is called the option grantor or writer. Direct costs involved in option trading are the cost of the premium plus the broker's commission.

In options trading, the option is taken out to conduct a specific transaction of a futures contract. The specific futures contract that can be bought or sold is called the underlying futures contract. The price at which the futures contract can be bought or sold at any time prior to the expiration date is called the strike price. The money the option buyer pays for the right to buy or sell a defined futures contract is the premium.

A commodity put option (right to sell) for September would give the buyer the right to sell a September commodity futures contract at a specified price, before the expiration date – regardless of what the commodity futures price might be at the time the option is exercised. A farmer could get some protection against the effects of prices falling by buying an option to sell a futures contract. For example, sometime before shearing, a woolgrower could buy a put option (right to sell) granting the right to sell a wool futures contract at a specified price that the woolgrower regards as satisfactory. If, after shearing, wool prices have declined below the level of earlier expectations, the farmer can exercise the put option (right to sell) and sell a futures contract for the previously specified price. Then, on selling the wool, the futures contract will be closed out by buying a futures contract at the lower price. The planned-for price is thus achieved. If, however, wool prices had risen above the earlier expected level, then the farmer would not exercise the right to sell (put option) a futures contract, and would simply sell the wool at the higher than planned-for price. The premium paid on the put option is then seen, not as a loss, but merely as the price that had to be paid to get some protection against a fall in the wool price. The advantage of taking out the option, as against simply selling a futures contract in the first place, is that with the futures contract the farmer, while 'locking in' to a guaranteed price, is at the same time 'locking out' of the chance to get a higher price than this guaranteed price if wool prices rise. Note that the great majority of options that are purchased are not exercised in practice.

Thus options can be used to avoid the effects of a fall in price, while leaving open the possibility of receiving a higher price if prices rise – all done for the cost of taking out the option. The maximum cost of options to the buyer is the premium. As well, buyers of options are not subject to margin calls, as happens with futures contracts.

The call option (right to buy) is used to protect against a price increase. A good example of using a call option (right to buy) is the case of a cattle feed-lot operator who has to regularly buy grain for use in fattening stock. The feed-lot operator buys the right to buy a grain futures contract of a specified type, for a specified price, within a specified time. If the price of grain rises later, the lot-feed operator can buy the futures contract, then buy the higher-priced grain and sell a futures contract at a higher price. The gain made on the futures transactions will roughly offset the higher than planned-for price paid for the grain. The overall position will be that the grain has been acquired at close to the price that had been originally intended. If it happened that the grain price had fallen by the time the feed-lot operator wanted it, then the call (right to buy) option would not be used, and the cheaper than anticipated grain would simply be purchased. The option premium is then the cost of guaranteeing a maximum cost of grain. The strength of using options, over futures trading without options, is that the option buyer faces a known and limited cost. If prices do change, the option buyer will either reap the benefits of a price rise or will make the same profit, minus the option premium, as would be achieved by solely using futures.

There is a further finesse to the use of options if an opportunity arises. Options can be bought to exploit a changed situation. Suppose a forward contract has been made to sell a commodity at a certain price. Then, the price rises. This means a potential 'loss' for the commodity seller, as greater returns could be achieved if the produce wasn't already sold forward. The producer can buy a call option (right to buy a futures contract) at the now higher price. If the commodity and futures price continues to rise, then the call option (right to buy) can be exercised. The futures contract will be bought, then later sold, at the higher price, thus making a profit on futures. If it happened that the futures price didn't rise above the strike price of the call (right to buy) option, then the option wouldn't be exercised and the option buyer would be worse off by the amount of the premium. But the buyer of the right to buy option was in with a chance to participate in the rising prices if they had continued. Similarly, put options (right to sell a futures contract) can be purchased to make a profit from an expected fall in price. At the expiration date, if commodity prices have fallen, and so too have futures prices, then the right to sell a futures contract will be exercised, and a futures contract will be bought back at the now lower price; thus a profit on futures trading will be possible. A consequence of the opportunities for using options in these ways is that options have a time value, which varies according to their potential value if exercised.

The seller of an option is paid a premium and in effect takes the opposite position to the buyer of an option. If a seller of an option sells a call option (right to sell a futures contract) and it is exercised, the buyer of the option will take out a futures contract to sell the commodity. The option seller is obligated to assume an opposite position and takes out a futures contract to buy (called 'going long'). This is how the risk is shifted. As with futures, for every winner there is a loser. If the option holder is going to exercise their option right, and thus make a profit on the futures transactions (which is how they receive, or pay, the price they had initially wanted), then the option granter will incur an equivalent loss on futures trading by assuming the opposite position. So, obviously, option sellers are punters who are betting that the undesirable events feared by the option buyer won't eventuate. Or, option sellers are betting that the loss they incur on the futures transaction will be less than the premium for which they sold the option. An option buyer (or seller) can cancel out their option by selling (or buying) an identical but opposite option before the expiry date of the option.

The price that is paid for an option depends to an extent on the strike price. That is, the price of an option depends on the amount of insurance that is offered by the strike price of an option. The less the amount of 'insurance', the less the price. Option prices are determined by buyers and sellers competing with each other to buy and sell 'insurance'. Price is established by open competition in the same manner as futures contract prices are established. The main determinants of the size of option premiums are:

- the option strike price in relation to the price of the underlying futures contract;
- the length of time in which the option is valid; and
- volatility of the underlying prices – that is, the chance of it being exercised.

An option with a long period of time remaining until expiration will have a higher premium than an option with a short period of time until expiration. This is because the longer the time involved, the greater the uncertainty and scope for big price changes, and so the greater the amount of 'insurance' the option will represent. The same applies to an option with a strike price that is markedly different from the relevant futures price. In this situation, it is more likely that the option to buy or sell a futures contract will be taken up – thus the premium asked by the seller will be higher than otherwise would be the case. Similarly, the more volatile the market, the higher the premiums sellers will charge when granting an option.

## THE ROLE OF FUTURES MARKET INSTRUMENTS IN FARM BUSINESS MANAGEMENT OF PRICE RISK

Only a small proportion of farmers use futures markets, with the most activity being among cotton and some wheat producers. Benefits often attributed to futures trading include a more stable income, price 'insurance' and more market information; and from these benefits derive further advantages of better budgeting and planning, and better access to finance. These are all valid and achievable benefits. The extent to which these benefits are achieved depends on a multitude of other factors. An important point, though, is that farmers can reap the benefits of futures markets as long as other businesses in their marketing chain use futures trading in a beneficial way to manage risk. It may be that trading of other instruments – in particular, exchange rate futures – is the main means to reduce exposure to price risk. Production risk affecting quantities of output produced discourages some farmers from committing to forward and futures pricing arrangements. Regardless, commodity producers are speculators until their product is sold, unless the producer selects a future price that is deemed to be satisfactory and acts to secure that price for the product – with the proviso that the premium or discount in relation to the contract price must be allowed for. Another view is that if the grower is committed to speculating in agricultural product prices, then it is easier and cheaper to do so using futures than by offering product for sale and then withdrawing it in the belief that a better price could be obtained at some later time.

A benefit of using futures ought to be more certainty about prices received and less variability of those prices over time. The worst effects of unexpected price falls might thus be avoided. An advantage of futures dealings rests in the woolgrower being able to choose the time to accept the judgment of the market about prices. Perhaps the main advantage lies in the opportunity that advance sales of contracts provide for budgeting for cash receipts. There is no reason to believe that the agricultural producer using futures will receive higher prices on average from using futures sales than from not doing so (apart from having sheer good luck) unless he or she is a skilful speculator (and if so, they might be better off doing this than farming).

Wool, beef and wheat are products that vary over a great range of grades. Uncertainty about the relationship between a particular cash price and the futures deliverable grades is called type basis risk. If the commodity doesn't fit the contract specifications, then speculation comes into the operation. Futures users are speculating about the difference between the contract futures price and the physical price of your commodity. There is an unavoidable need to estimate the discount/premium of the product compared with the specified futures contract.

Another reason for farmer reluctance to use futures is their interpretation of futures prices. If futures prices are below spot prices (an inverse carrying charge), this indicates a current shortage of supply and a signal to reduce inventories. A discount of futures prices over cash prices doesn't indicate that futures prices must rise. If people in the market expected rising price levels, futures buying would cause futures prices to rise. Alternatively, a situation in which futures prices are at a premium over spot prices reflects a current surplus of supply, and provides an incentive to carry stocks forward. The relationship among spot and various futures prices is a guide to the merchant or user about what to do with stocks. They don't represent forecasts of future cash prices. However, cash and futures prices converge as the delivery month approaches.

A common misinterpretation is that the movement in cash and futures prices permits a routine hedge in which the profit from one position always offsets the loss from the opposite position. The convergence of cash and futures prices, caused by the threat of arbitrage, makes it possible to offset to a considerable extent outcomes on futures and in cash markets. However, for practical reasons, convergence in the delivery month isn't 'perfect'; therefore, the 'routine' hedge isn't quite as routine as it is often made out to be.

There are many reasons, though, other than simple lack of understanding, as to why futures have not been widely used by farmers. A lack of speculative capital in commodity futures markets can be a cause of limited activity – and the limited activity can be a cause of a lack of speculative capital. The development of 'over the counter' products also could limit the liquidity available for futures trading. From a farmer's viewpoint, a main reason for not using futures trading might be that there are many alternative ways of reducing risk, and producers have traditionally preferred low gearing, enterprise diversification or specialisation, and investment strategies on and off the farm, to futures trading.

A habit of seeing losses on futures as losses in total and not as, say, the price of reducing risk (hedge to lose), or at least as being the same as the opportunity losses often incurred by any too early or too late product trading decisions, is another reason for producer reluctance to trade on futures markets. Furthermore, primary producers in Australia are acutely aware of the difference between price risk and income risk. That is, output fluctuations play a large part in income fluctuations, and reducing price fluctuations can go only a part of the way to stabilising incomes.

The ability of a business to service debt and to grow depends on prices received, quantities (and quality) produced, and total fixed and variable costs. While fluctuating prices can be one source of risk, and contribute to a business going bankrupt, fluctuations in quantity of output produced and the level of fixed costs relative to total output are likely to be bigger contributors to farm liquidity problems and to bankruptcy. Price risk management strategies can help to reduce the variability of cash flows. Farmers can sell some of the price risk they face to speculators by hedging the prices they will receive, using futures markets or options. Efforts to stabilise prices received won't solve a fundamental structural problem with a farm business, such as the overheads being too high relative to output, or gearing being too high relative to the expected mean and variance of expected cash flows. Using futures markets cannot affect the absolute magnitude of the net cash flows each year or over a run of years. If a farm business is struggling economically because of low net income, or struggling financially because of high gearing, greater price stability resulting from using commodity futures won't turn the situation around. Indeed, even for farms with a good future, reducing price risks via futures and options trading isn't likely to be the main tool of financial management. Futures, if used at all, will be tools that will complement the many other risk management actions that farms with good prospects will, almost by definition, be doing already.

There are a number of points to note about futures markets:

- Agricultural commodity prices reflect the fundamental forces in the market, and futures market operations cannot raise prices.
- Using futures cannot do anything for a business if the real difficulties arise from fundamental structural problems in the business.
- Farmers take many steps to manage all the risks they face.
- A well-functioning futures market could be of significant benefit to farmers, even though most might never trade a contract, by facilitating more informed and efficient pricing and operations by others in the marketing chain.

The benefits of futures markets are diffuse. Futures markets in general can play a vital role in improving efficiency of pricing and operations throughout the production and marketing chain, and in facilitating other efficiencies and developments in production, financing and selling.

In the next chapter, the questions that are addressed concern the extent to which a farm business can profitably extend its comparative advantage further through the marketing chain.

## CHAPTER SUMMARY

*Risk and uncertainty are part of all business activities and actions and, importantly, risk creates rewards. Above-average returns can only be reaped if the investor is willing to take above-average risks.*

## REVIEW QUESTIONS

1 Give examples of sensitivity testing, scenario analysis and breakeven analysis.

2 It is higher risk that creates the opportunities for higher profit. To make higher-than-average profits, investors must take higher-than-average risk. Explain.

3 A futures market transaction is a temporary substitute for a subsequent physical transaction. Continually going back to this statement is the key to understanding who does what, when and why in futures trading. Give an example of a futures market transaction that might be undertaken by (a) a wheat grower, (b) a flour miller, and (c) a speculator who reckons

he has superior information about wheat supply and demand sometime in the future.

4 Options are an attractive option for hedgers because they make it possible to protect against adverse price risks without having to be 'locked out' of unexpected favourable price outcomes. Explain how an agricultural commodity producer facing price risk could use options.

5 To work well, commodity futures markets need lots of buying and selling. There has to be a lot of risk to be shifted, and a lot of speculators willing to take the risk off people wishing to transfer the risk. That is, futures markets work well when there is a lot of action – and there is a lot of action when they work well. Do the Australian commodity futures markets have a lot of action? Why? Why not?

6 Risk is a commodity that can be sold by people who don't wish to bear it, to people who are willing to carry it because of the opportunities it creates to make profits. Discuss.

7 In futures trading, for every buyer there is a seller; for every winner, there is a loser; a lot of speculators lose a little and a few win a lot; and the system of margin deposits keeps everyone honest. How do margins work?

## FURTHER READING

Anderson, J.R. and Dillon, J.D. 1992, 'Risk analysis in dryland farming systems', FAO Farm Systems Management Series No. 2, Food and Agricultural Organization of the United Nations, Rome.

Arrow, K. 1992, 'I know a hawk from a handsaw', in M. Szenberg (ed.), *Eminent Economists: Their life and philosophies*, Cambridge University Press, Cambridge.

Bernstein, P.L. 1996, *Against the Gods: The remarkable story of risk*, John Wiley & Sons, New York.

Boehlje, M.D. and Eidman, V.R. 1983, *Farm Management*, John Wiley & Sons, New York.

Gigerenzer, G. 2002, *Reckoning with Risk: Learning to live with uncertainty*, Penguin Books, London.

Gigerenzer, G. and Todd, P.M. 1999, ABC Research Group, *Simple Heuristics That Make Us Smart*, Oxford University Press, Oxford.

Heady, E.O. 1952, *Economics of Agricultural Production and Resource Use*, Prentice Hall, New York.

Hopkin, J.A., Barry, P.J. and Baker, C.B. 1999, *Financial Management in Agriculture*, The Interstate Printers and Publishers Inc., Danville, Illinois.

Murray-Prior, R. and Wright, V. 2001, 'Influence of strategies and heuristics on farmers' response to change under uncertainty', *Australian Journal of Agricultural and Resource Economics*, vol. 45, pp. 573–98.

Pannell, D.J., Malcolm, L.R. and Kingwell, R.S. 2000, 'Are we risking too much? Perspectives on risk in farm modelling', *Agricultural Economics*, vol. 23, pp. 69–78.

Tomek, W.G. and Robinson, K.L. 2003, *Agricultural Product Prices*, 4th edn, Cornell University Press, Ithaca, New York.

Wright, V. 1983, 'Some bounds to the relevance of decision theory', *Australian Journal of Agricultural Economics*, vol. 27, pp. 221–30.

# 6 Marketing agricultural products

Farmers, and other agribusiness firms, naturally need to make decisions that influence the value of their output to their customers. Choices about when and how to sell output are some of these decisions. However, most of the main decisions affecting the value of output are made long before selling occurs. This chapter is about ways of thinking about the links between customer preferences and agribusiness management. Key questions that are considered in this chapter are: 'What is the role of my firm in the agribusiness system? How should this inform my management decisions? And what changes to the way I relate to other participants in the system might be feasible and profitable?'

As with our earlier chapters, a guiding assumption in this chapter is that meaningful analysis of a situation is fundamental to informed decision making that contributes to the achievement of objectives.

## THE AGRIBUSINESS FIRM IN ITS MARKETING ENVIRONMENT

We have so far emphasised the importance of analysis to reliable decision making on farms. Our focus has been on achieving objectives by the adroit employment of resources in an environment characterised by a good deal of uncertainty. In such an environment, managers who fail to define the determinants of performance and the nature of risk, and to respond accordingly, are effectively speculating and leaving the satisfaction of objectives to chance. It is essential to understand the rules of the game, the field on which it is being played and the manoeuvres that are valid.

Important strategic questions are to do with the choice of enterprises to consider as options, the way the manager views the farm in the context of the agribusiness system of which it is part, and ways in which the farmer may strive to modify the rules of the game.

In this chapter, we refer often to marketing systems. This is a generic label that includes what, in agricultural economics, we call agribusiness systems. They are the entire 'vertical slice' of an economy involved in the creation of a category of output for consumers in society at large. 'Market levels' within marketing systems are stages in such systems where changes in ownership (that is, exchange) occur.

Many farmers are captives of the highly competitive structure of agriculture in Australia: they think about farming with a strong on-farm emphasis, lamenting the reality of their lack of control over price, being overly receptive to those who offer panaceas for the reality that they lack control over the key determinants of financial performance, whether the so-called solutions are benchmarking or 'better marketing'. Accepting such nostrums indicates a lack of coherence in strategic thinking; not properly understanding the game being played.

In this chapter, we consider marketing in the context of agribusiness systems. Our purpose isn't to emulate mainstream business marketing texts. Instead, we have aimed to present conceptual frameworks that enable people running firms in agribusiness systems, including farms, to analyse how agribusiness systems work, in ways that help managers of agribusiness firms to define their own role in those systems, and help them to make decisions about their output accordingly. In this chapter, ways to analyse the business environment external to the firm are considered. Business marketing ideas are used because they sit at the heart of the business of agribusiness systems: the creation of output valued by consumers.

The term 'marketing' is commonly used to refer to activities, such as distribution, further transformation and promotion, that occur following the production of output. This is true, also, in mainstream agricultural economic literature where marketing is identified with various functions that add utility, or value for customers, to farm output – for example, time, form, and place utility. In mainstream (business) marketing literature, 'marketing' is defined formally to include, as well, decisions made about the characteristics of output itself. The rationale for this is that decisions about the precise nature of product to produce are ultimately a major part of the value that customers place on output. Post-production decisions, to do with distribution and promotion, can add to, or reduce, the value placed on the characteristics of the output as it is presented to consumers.

Typically, in agricultural economics, the focus in marketing tends to be on value that is added to farm output, often in an aggregate sense. The marketing focus in this chapter is on how an individual firm creates the total value that the consumer places on the output – that is, the value in the characteristics of the output, plus the value added by distribution and promotion, as a whole set of value traits. This firm could be a farm. The firm's customers could be at any of a number of marketing levels within the marketing system.

In agricultural economics, it is usually assumed that the price of farm output is formed in markets since the highly competitive structure of most agricultural sectors means that producers have little or no control over price. In non-agricultural marketing, an assumption that is usually made is that the firm has the ability to set price. For this reason, in non-agricultural marketing, price is another component of the 'whole set' of value-traits, that – like product, distribution, further transformation and promotion – can be selected and be more or less attractive to customers. Indeed, an aim of marketing activity is to make a product sufficiently different from the product of competitors so as to achieve some ability to set a price (that is, a price maker) different from the price that sellers of undifferentiated product face (that is, the price takers).

This 'set' of value-traits in the product – promotion, distribution and price – is called the marketing mix. The term 'marketing mix' refers to all the characteristics of the good or service that a firm presents to customers. A common description of this marketing mix is the 'Four Ps of marketing': product, promotion, place (distribution) and price. A central idea in marketing is that this marketing mix can be optimised for a particular good or service. The aim is to strike the best balance between (a) the needs and capabilities of the firm, (b) the marketing mixes of competing firms, and (c) the preferences of customers.

In this text the discussion of agricultural marketing is initially grounded in a focus on agribusiness systems, rather than on the individual firm or farm. We do this for several reasons. One reason is that the highly competitive structure of farm sectors raises fundamental issues to do with marketing. These issues cannot be understood properly without considering system-wide questions. Another important reason for focusing on agribusiness systems and not individual firms is that when emphasising the individual firms responsible for creating the core product within a system, there is a tendency to poorly conceptualise the roles of *other* firms in the system. Finally, the intrinsically strategic issues related to relationships between firms and farms can only properly be analysed in a system context.

## MARKETING DEFINED

Historically, 'marketing' meant 'shopping': 'doing the marketing' meant 'doing the shopping' – going to the market. As a business discipline it embodies the same central idea: marketing means identifying objectives to be satisfied through exchange processes, and going to a market to seek to satisfy these objectives. The main distinction between consumers and producing firms going to market is that one seeks to acquire products (goods and services) and the other seeks to sell products.

Until the twentieth century, producing organisations tended to regard selling and distribution activity as being quite separate from production. It steadily became apparent, as selling was analysed more closely, that many characteristics of the output being sold were critical to the ease with which sales could be achieved. This means that, taking a medium-term view, the sophistication of selling and distribution activity is of marginal importance to sales, compared to the valuation consumers place on the characteristics of products. Recognition of the importance of the value consumers place on the characteristics of the product has led to awareness of the fact that 'marketing' for firms therefore also includes decisions about what is produced. That is, marketing concerns what bundle of characteristics are being produced, as well as what other desirable characteristics are being added, to try and keep customers happy.

At the heart of the process of firms deliberately contemplating ways to make profit through exchange is the choice of what products to produce. Marketing starts with decisions about what to produce. This seems obvious. Nevertheless, it is still commonplace for people running firms, including farms, to ponder how they can improve the performance of their business with 'better marketing', when the fundamental question of the primary attractiveness of the characteristics of the output of their business to customers is far from their minds.

Two factors encourage the approach to marketing as an 'add on' to the product, instead of including choices about the product itself as being part of decisions about marketing. One factor is that changing output is usually a much more significant exercise than changing promotion, for example, and people running firms may be uncertain as to what changes to output would be best, anyway. Another factor is the widespread myth that customers are gullible and can be easily swayed by clever promotional activity. That is: 'I'll produce what I want to produce and somehow easily persuade customers that this is what they need or want.' Belief in this myth reflects a poor understanding of consumer psychology and, more importantly, a short-term view that fails to account for the likely reaction of customers when the performance of the product fails to meet the expectations created by the 'clever' promotional activities.

Information is the core of marketing activity. Marketing by a firm implies that managers need to consider any and all of their decisions that may impact on judgments customers form about their marketing mix, compared to mixes presented by their competitors. Since all effort by firms can be seen as 'production', we can take the view that marketing is the series of activities that is intended to marry customer preferences and the firm's production. Then, optimisation of the marketing mix is the optimisation of production, where 'production' is defined comprehensively to include all of the characteristics of a firm's activity that are put before consumers.

So, the evolution of marketing as a discipline has led to explicit recognition of the fact that anything and everything a firm does may impact on sales and should, ideally, be undertaken with a sound understanding of relevant customer preferences.

## APPLYING MARKETING

Marketing theory has implicit assumptions that limit the extent to which marketing principles can be generalised to firms or situations. Specifically, all the pertinent components of marketing mixes – product, promotion, place (distribution) and price are assumed to be under the control of the firm. A central idea in marketing is that this marketing mix can be optimised for a particular good or service. That is, as noted above, the aim is to strike the best balance between (a) the needs and capabilities of the firm, (b) competing firms' marketing mixes, and (c) the preferences of customers. These situations – control of all the pertinent components of the mix, and the potential to optimise this mix – is by no means common, and is especially uncommon where farms are concerned.

Another assumption about marketing is that the part or level of an agribusiness or marketing system where a firm's customers 'go to market' is always apparent. Whether a firm should define its customers as the final, retail-level customers, or as some 'organisational' market in between, is often assumed to be obvious. This isn't necessarily so, and since marketing is about orientating productive efforts to consumer preferences, the definition of who are customers of a firm is a significant question. Time and effort put into identifying who is the customer is time and effort well spent. It is the starting point of marketing activity.

A further assumption that has developed with the evolution of the marketing mix notion is that, since any component, or combination of components, of the product, promotion, place and price mix may be central to customer choice, these components of the mix are equally important. This is rarely so. It ignores basic distinctions in the roles of the components of the mix: product, promotion, place and price.

In the following section, the marketing mix is described in some detail. The questions of the extent of control a firm has over the mix, and the relevance of components of the mix, are used to explain an approach to marketing that is relevant to agribusiness systems.

## THE MARKETING MIX

The marketing mix is a simple notion that denotes a complex reality: there are many dimensions of a firm's output. Inevitably, the product promotion, place and price components of the marketing mix overlap somewhat. This doesn't matter. It is important, though, to understand the essence of each component of the mix, because for some components – such as price, for instance – things might not be quite what they seem.

### Components of the marketing mix

#### *Product*

'Product' refers to all dimensions of the good, service, or combination thereof, that a firm produces. This has both static and dynamic elements. It includes the set of characteristics of the product that help to satisfy what are called basic motivation or 'trigger needs' of the consumer – for example, the beans in a can of baked beans, or the security and interest rate of a banking service. These are characteristics of a good or service that offer satisfaction of the basic needs that 'triggered' the move by customers into the market for these goods or services.

The product component of the marketing mix also includes other features that surround the characteristics that meet the trigger needs. These other characteristics may appeal to further needs and preferences of customers, other than the basic 'trigger need'. For example, the packaging of the baked beans, or the friendliness and courtesy of bank service providers, are part of the product, but they don't normally provoke (trigger) the initial movement of customers to seek the product. These additional characteristics of the good or service are likely to influence the choice of providers, though, once the customer has decided to seek to fill a 'trigger need' and seek out baked beans or a bank's services.

So, if we put ourselves in the customer's situation, the need or anticipated need for food (the trigger need) prompts the customer to put baked beans on their shopping list. Once on the list, the question arises as to which brand of baked beans will be purchased. Other characteristics of the product may be pertinent in this choice process. The sizes of product available, the visual appeal of the packaging, the ease of opening of the can – these are all characteristics that are likely to influence choice. Each of these is called a search characteristic; customers can detect the presence of these characteristics merely by inspecting (searching) the product.

There are other characteristics of product that are not detectable by simple inspection. The flavour of the beans in this can of baked beans remains an unknown, as does the risk of food poisoning from consuming them. These characteristics – flavour and risk – are experience characteristics. The presence or absence of experience characteristics can only be confirmed by consuming the product. The presence or absence of experience characteristics may create risk in the customer's mind: the risk of an inappropriate choice. That is, the risk might be created in the mind of the customer that 'these beans might poison me and my family, but I won't know until we have eaten the beans and they have poisoned us'.

A third category of product characteristic is credence characteristics. The customer can never readily confirm personally whether or not these types of characteristic actually are characteristics of the product. Credence characteristics include claims about production methods (dolphin-friendly tuna, organically grown vegetables, freedom from genetically modified (GM) inputs) or other producer behaviour (percentages of revenue given to charity, environmental sensitivity). These are also possible sources of perceived risk. For example, someone might consume the product because they think these presumed credence characteristics are real, and they may turn out to be illusory.

The experience and credence characteristics of a product can create doubt and risk when customers place value on such characteristics. If experience and credence characteristics are of interest to customers, and are significant criteria for choosing between products, then not being able to know for certain about the experience or credence characteristics at the time of purchase makes the purchase risky.

There are various ways customers can alleviate the risk they perceive about the experience and credence characteristics of a product. One way is to rely on a third party as a provider of an implicit or explicit guarantee about the characteristic in doubt. If concerned about food safety, for example, a customer may rely on the conviction that their retailer only offers safe products, or that 'the government' oversees properly such issues. Customers may rely on accrediting authorities as to the healthiness of a product – for example, the GM status or organic status of foods.

Another response of consumers to risk associated with characteristics of a product is to engage in search activity, such as talking to friends or colleagues or checking consumer organisation information. (Contrary to the popular view of the susceptibility to advertising and other forms of promotion, it is difficult to overstate the extent to which final, retail-level customers rely on friends and colleagues for choice-related information about products. They have credibility that producers can rarely match.)

A third response to risk associated with a purchase of a good or service is to rely on a brand. A brand is useful to the extent that it is perceived to offer a reliable promise of the presence of choice-relevant characteristics that are experience or credence characteristics. A brand offers little if search characteristics alone are the bases of choice. That is, the characteristics we can see and feel are the criteria for choice and we are not concerned about experience or credence risks. When a customer feels that they can rely on a brand to deliver experience or credence characteristics, the brand itself becomes a search characteristic.

Thus, brand is a component of the product within the marketing mix. Many products are branded. Whether or not the brand has value for customers depends on its ability to reduce the risk they perceive in choosing products. This, naturally, depends, in turn, on the reliability customers perceive in the delivery of experience or credence characteristics by products bearing that brand. This is called brand integrity. The key implication, which can be highly significant in agricultural marketing systems, is that control over the various characteristics relevant to consumers has to be available to a sufficient degree. The central criterion in forming a judgment about the appropriate investment by farm firms in their marketing mix is the extent to which they have control over the various components that are relevant to consumers.

### *Promotion*

'Promotion' refers to all communication by the firm that is intended to reach customers. This includes advertising (which involves the use of mass media of any kind to put specific information, designed by or for the firm, before the public), personal selling (meaning the activity, by representatives of the firm, of interacting one-to-one with customers), public relations (the management of information released to the media, the use of which isn't under the control of the firm (unlike advertising)) and sales promotion. Sales promotion is a catch-all category of promotion which includes all remaining forms of promotional effort. Point-of-sale material (brochures, leaflets, in-store banners), mail-outs, free trials of products, price discounting, 'frequent' user schemes and quantity discounts are all examples of sales promotion.

Promotion involves two main intentions of a firm: the provision of information designed to help customers make choices (favourable to the promoter, hopefully); and the 'pushing' of a product by creating incentives to purchase, including what is called, in law, puffery – emphasising the positive features of a brand.

### *Place*

Place, or distribution, involves the processes by which a product is moved from producer to customer or the market. It includes processes of warehousing and choice of intermediaries. The part of the entire marketing system between a producing firm and their customers is called the marketing channel.

### *Price*

Price refers (only) to the list price the customer pays for a product. It doesn't refer to the economic cost of acquiring a product. The reason for this is that the marketing mix is the set of decisions made, and costs incurred, by the firm, rather than the unique economic reality of, and acquisition costs borne by, customers. So, price doesn't include costs the customer might incur in getting to the shops, interest on credit card debt, opportunity costs, or the costs of storing the product. Nor does price include promotional price discounts offered to customers – these are seen as variations on set prices that have a promotional intent and are thus an aspect of promotion.

### *Conclusion*

Decisions made about the marketing mix ideally have two features. First, decisions need to reflect customer responsiveness to various values/levels of the product, promotion, place and price components of the marketing mix. Hence, it is critically important to know about the preferences of customers. Second, decisions should be thematically consistent. Because all components of the marketing mix contain information, it is desirable that the information is consistent. For example, the nature of packaging, the distributors used, the promotional propositions made and the price set can all signal a certain quality position of a product in a market. These decisions should be such that each gives off the same signals. Competing signals will reduce the overall impact of the marketing mix, undermining its effectiveness.

## Managing the marketing mix

As noted above, management involves identifying the appropriate determinants of the performance of businesses on which to place most emphasis. Marketing management is no different. In marketing management the key issue is to do with customer choice. *The aspects of the marketing mix that warrant greatest attention are those that relate to the criteria that customers are using in making their choices.*

The choice criteria of consumers may exist in any of the various parts of the mix. Obviously, marketing mixes have many characteristics. Which characteristics are customers relying on to choose between alternative solutions to their needs? To identify which components of the mix that customers are actually using to make decisions, it is necessary to understand how customers perceive the choices confronting them.

Assume, initially, that a customer is encountering the need for food for the first time. The customer sees a world of alternative foods. Many product characteristics of foods serve only to define them into the broad category of 'food'. Products without such characteristics are not considered as relevant to the need for food.

As the customer trawls through all the possible foods that might be acquired to meet this need, many criteria are used to eliminate alternatives. (This is called an elimination by aspects choice model.) Some foods will be culturally unacceptable, too expensive or too hard to prepare to eat; they may have an unattractive appearance or flavour; be impossible to buy; be of unreliable or unknown quality; be non-organic; and so on. Ultimately, the customer will arrive at a subset of foods that are alternatives worthy of close consideration at this particular time. This is called the consideration set or evoked set, which means the grouping of products that are likely to meet the needs of the customers. Foods that are not in this grouping will be given no further consideration and will be excluded from the potential purchases.

The food that is selected will be the food that is perceived by the customer as the best choice on the basis of the few criteria they are using to select among the alternatives being considered. For the producer, the key questions are: 'What are these criteria on which the customer will make their choice?' and 'How do the characteristics of my product compare to the other alternatives in the grouping of products the consumer has identified as being likely to meet their wishes?' – that is, the consumer's consideration set.

Characteristics that alternative products are perceived to possess in equal quantities will be useless as choice criteria. Choice will be based on perceived differences, not similarities. These differences may or may not be directly related to the trigger need, the need for food. All the products in the consideration set will satisfy the trigger need.

The marketing mix can be divided into two categories of characteristic: those that help to satisfy the trigger need, and those that facilitate exchange. Product is usually in the category 'contributes to satisfaction of the trigger need', while promotion, place and price are usually in the category 'characteristics that facilitate exchange'. Choice criteria are likely to come from either category. Flavour may govern choice, or ease of purchase may be the over-riding consideration in the decision. However, both characteristics are included in the 'weighing up' or decision process. Trade-offs between attributes are involved. Ease of preparation may govern choice, or price may decide. Ultimately the decision, or choice, will depend on customer preferences and the similarities and differences between alternatives within the consideration set.

Components of the marketing mix that facilitate exchange are important influences on choice. However, in the medium term, product counts most. *It is the capacity of a product to satisfy trigger needs that makes it the focus of exchange over time.* The important point that follows is that changes to product are the changes to the marketing mix of a firm that offer the most enduring possibilities for a firm to differentiate its product from competitors' products. Product is also the element of the mix, unlike the other components, that has the potential to 'travel down' a marketing system. That is, change to the characteristics of product can be implemented at various levels in the marketing system, not *just* at farm level or even *necessarily* at farm level. This is important in the context of the discussion below about vertical segmentation in marketing systems.

## DEFINING CUSTOMERS

Defining those characteristics of the marketing mix that matter most involves defining who are the customers of a firm. In the mainstream marketing literature, 'market segmentation' and 'targeting' are discussed as the ways to define the customers. This involves breaking down a given market into groups of customers with relevant similarities; most notably, that they apply the same criteria for choosing a product category and hold similar preferences about those criteria. These 'segments' of markets are groupings of customers who can be identified as seeking a package of marketing mixes that are different from the mixes sought by other customers in other segments of the market. This process of breaking consumers into like groups can be done at any market level. We can consider segmentation of final customers, or of intermediate market levels such as processors, wholesalers or retailers.

If we consider a simplified model of a beef agribusiness system (Figure 6.1), we can illustrate some of the ideas here. The bases for market segmentation at each market level are differences among customers of the producers supplying that level. These differences may be associated with differences in preferences that make it possible and profitable for suppliers to target groups of customers with different marketing mixes. Thus, input suppliers may present different versions of products (including size or volume differences), different sales force approaches and different price schedules (including promotional quantity discounts) to large, corporate specialist farmers than to small, family specialist farmers. Likewise, wholesalers may profitably present different mixes to supermarkets than to the food service sector (restaurants and fast-food outlets).

Once a market has been split into groups of customers with similar preferences, a firm can consider whether, and to what extent, it would be sensible to aim at, or 'target', specific segments. What is involved is moving away from a single marketing mix that offers identical characteristics to all customers, to a variety of mixes that more closely meets preferences that differ from one group of customer to another. Naturally, to do so, relevant components of the marketing mix have to be under the control of the firm. Whether a segment is worth aiming at depends on a number of factors. The consumers in the segment have to be

**Figure 6.1** Beef agribusiness system

willing to pay more than the marginal cost to the producer of tailoring the marketing mix to their preferences. These costs can be considerable. Targeting segments, rather than producing a single product line, commonly implies smaller production 'runs' and the loss of economies of scale. Marketing (including production) activities are multiplied.

The targeted segments have to be genuine potential sources of increased sales and profit. There also has to be a reasonable prospect of the segments being stable, over time, since the firm needs to commit resources to targeting and will require returns over a time period. It may therefore be appropriate for a firm to target some segments and not others. The merit of investing in marketing activity has to be judged like any other investment.

If competing firms are segmenting a market, there may be little choice for a producer other than to do likewise. Since consumer segments are defined on choice criteria, a failure to offer a product as targeted as those of competitors may cause a brand to fail to be considered by potential customers, or to be seriously uncompetitive.

## Market level

Segmentation and targeting are processes that can apply at any market level, not simply the final retail level. This process is called horizontal segmentation. The requisite first step is to identify *which* markets (that is, market levels) in a marketing system should be viewed as potentially including customers of a firm. This isn't necessarily always obvious. First, farmers, especially, are periodically admonished to be market-oriented: to 'think about the final consumer's needs and preferences'. The implication is that farmer decisions would be usefully, and profitably, informed by knowing a lot about the needs and preferences of the final consumers.

Another reason for identifying carefully the appropriate market level where a firm's customers may be is that in agribusiness systems and other marketing systems since the 1960s there has been the emergence of so-called generics: products with no brand. 'House brands' of retailers aren't quite the same thing, because they carry the retailer's brand. But both generics and house brands do share one effect: the producer who supplies them cannot sensibly regard final customers as being their market. The products which farmers supply to wholesalers and retailers as house brands or generic, no-name products represent a case where the market of the farmer is the wholesaler and retailer. The producer who produces both no-name products and branded products has two market levels to deal with – one for branded output and one for unbranded, or anonymous, output.

A further consideration is that, in mainstream marketing literature, it is common to describe the firms between a producer and the final customer as the marketing channel. This is a series of firms through whose hands a product moves on its way to the final customer. Commonly, decisions about which channel to use are expected to be made on the basis of which channel will suit final customers best. Often, this is somewhat unreal. Intermediaries are not always simply distributors of a firm's product. Often they have considerable bargaining power in a marketing system. So, for many firms, it is necessary in practice for the firm to design marketing mixes with multiple market levels in mind. And, sometimes, these market levels may have conflicting preferences.

The process of identifying which market level, or levels, it is appropriate to define as the customers of the firm is called vertical segmentation. As with horizontal segmentation (aiming at customers at a defined level of the market), an important question with vertical segmentation (identifying the apt market level) when identifying customers is whether all levels that are identified should be targeted. We shall consider this question after we define what are relevant market levels.

We need criteria that enable us to determine which market levels are and are not customers of the firm. Otherwise, we don't have an informed basis for segmentation and targeting. These criteria flow naturally from the essence of marketing: designing production to optimise its appeal to customers. 'Customers' can only include those in the marketing system who have preferences that relate to the activities of the firm. 'Designing production' requires that the aspects of production that relate to those preferences are under the control of the firm. 'Optimising production' requires that there is the possibility of modifying revenue, as well as costs, by targeting customers.

The value of everything that occurs in a marketing system is determined, initially, by the response of the final customers to the output of the system. For this reason, when striving to define who might be the customers of a firm, we might begin with the objective of defining the final customer as the preferred market level.

The first criterion that must be satisfied, for customers at a market level to be possible customers of a firm, is that the customers are responsive to changes in the attributes that comprise the marketing mix. If customers are indifferent to changes in the marketing mix of the firm, for whatever reason, they cannot qualify as customers of the firm who are worth identifying. Note that this is never the case at the first market level the firm serves. At the very least there is always some degree of responsiveness to variations in price and product quality at the first market level. The question is: 'How far through the marketing system does responsiveness show up?'

The component of the marketing mix that attracts a response at the first market level a firm encounters could be any part of the mix. It could be product quality, promotion, distribution or price, for example. The exchange facilitation aspects of the marketing mix (quality, promotion, distribution), however, are normally ephemeral in their impact on customers; the demand for these aspects is met at the first point of exchange, the immediate market. Product is the mix element that contains characteristics that *may* traverse multiple market levels. Promotion, to the extent that it is informing customers of valued product characteristics, may also span market levels.

There is an important point to note here. It is a reasonable thing to suggest that *all* firms in a marketing system are contributing to the marketing mix presented to the final customer. *This doesn't imply, however, that the final customer is responsive to changes in the mix of any single firm in the system apart from the nearest to them (usually a retailer). It is a question of fact. It will depend on the extent to which a firm's mix characteristics continue to be present in the mix being assembled by the marketing system, and the extent to which those characteristics are choice criteria for the final customer market or a segment within it.* In terms of managing the marketing mix to optimise its appeal to customers, this is plainly a meaningless objective for markets that are indifferent to variations in the firm's mix.

## Control over characteristics of components of the marketing mix

Once a firm has identified the various market levels that are responsive to changes in the marketing mix, the second criterion applies: are the relevant components of the mix under the firm's control? This is an issue, for two possible reasons. One is that there may be reasons at the level of production that deny control over characteristics. Seasonal variability, or drought, for example, may vitiate intended levels of beef quality or wool fibre strength. The other reason is that a characteristic of value to customers, such as meat tenderness, may be affected by the actions of other firms (through rough livestock handling and inappropriate slaughter techniques), and beyond the control of the firm that produced the initial product. Whatever the reason, if a characteristic that matters to customers isn't under the control of the firm, it makes little sense to contemplate trying to optimise its status for the customer. Where this is recognised as an issue, it may make sense to consider ways of modifying the marketing system to achieve control. This is discussed later under 'Strategic marketing'.

## Premiums for better marketing mixes and who gets them

If a firm has defined customers in markets, other than the first one it encounters in the marketing system, the final criterion to confirm their status as being genuine customers of the firm is that premiums they will pay for better mixes from the firm will actually return to the firm. Beyond it being possible to modify the mix in ways that customers like, there is no point in doing so unless the positive sales volume and/or price responses of the customers flow back, to an adequate degree, to the firm. This may not be the case if intermediary firms are able to deliberately absorb some of the benefits; or if the competitive structure of intermediate markets inhibits the flow of benefits back to farmers; or if the benefits attract prompt increases in matching output from competitors that have the effect of reducing the benefits after a short time.

## Outcome of analysing market level

The simplest outcome of the analysis of market levels is that there is only one market level composed of customers of the firm. The most complex possible outcome is a set of different market levels that are customers of the firm.

It is important to understand what these outcomes mean. When there is a single relevant market level, it is appropriate for the firm to focus only on that level in its mix management; all output decisions should be informed by the preferences of, and competition for, customers at that single market level.

When there are multiple market levels, it means that deliberate contemplation of preferences and competition at different market levels is appropriate. In conventional economic analysis we take the view that, if there is an acceptable level of contestability in intermediate markets, preferences from lower markets in a marketing system will be reflected in relative prices paid to firms supplying products to customers at the intermediate market levels. That is, since prices reflect all relevant preferences in the agribusiness system, there is no need to investigate preferences directly.

However, the source of innovation in characteristics coming from our focal firm is that firm. If a firm seeks to differentiate itself from competitors, and/or to extract more profit from its resource use, innovation in its productive activity is required. Where this innovation relates to characteristics valued by more distant markets than its immediate market, understanding those markets is key. Relative prices only reflect the valuation of existing output and may or may not indicate desirable innovation. Only in the situation where a firm can rely on its immediate market to provide information, in addition to relative prices, about the preferences of customers in lower relevant markets will it not be necessary for the firm to analyse the preferences directly.

Even in marketing systems operating well from an economic, competitive perspective, full transmission through different market levels of information about the preferences of customers is by no means assured. The operators of intermediate firms would need to judge that they have a vested interest in enhancing the value of the output of the initial supplying firms – that is, there is something in it for them. Otherwise, they lack an incentive to gather or pass on the information required for meaningful innovation to the marketing mix of the initial supplying firm.

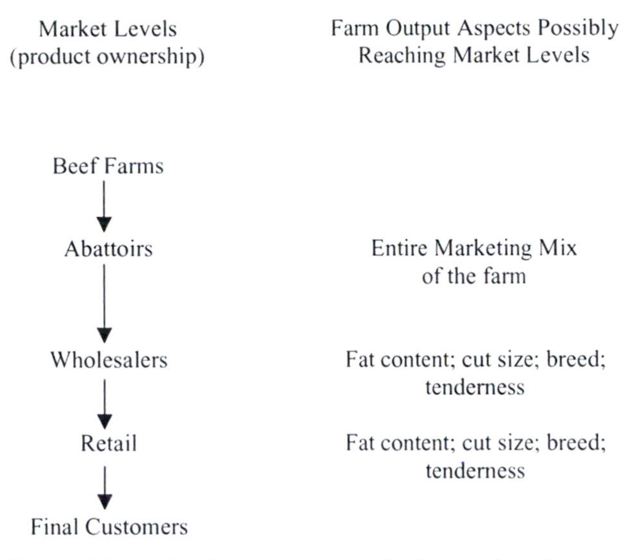


**Figure 6.2** Vertical segmentation in the beef agribusiness system

We can illustrate some of these ideas with reference to Figure 6.2. This figure identifies the possible flow of farm output characteristics that *could* pass through the beef agribusiness system. Initially, the first market level (abattoirs) is presented with the entire marketing mix of a farm. As the farm product undergoes further transformation there are several characteristics that persist in the product: fat content, cut size (influenced by size of beast), breed (defining flavour) and tenderness. There are other characteristics that might also flow: organic production, a farm brand, a specific production location.

If one or more of these characteristics that flow through the system (a) matter to customers at a market level, (b) are controllable by the farmer, and (c) can attract a premium that flows, at least in part, back to the farmer, then we have the preconditions for a farmer targeting a variety of market levels. If any of these criteria are not satisfied at a market level, it is nonsense for the farmer to imagine that customers at that market level are, in any sense, 'theirs'.

## STRATEGIC MARKETING

When we discuss innovation to the marketing mix and analysing a system to define vertical segments of marketing systems, we are in the domain of thinking about strategy. This is because we are considering information that will be used to inform infrequent and long-lived decisions, requiring substantial commitment by the firm.

There are two types of strategic decision involved. The more obvious type is innovation in the marketing mix. Here we are referring to changes to mix characteristics in order to aim at new market levels. In the agribusiness context, for example, this would be exemplified by a decision to produce larger, leaner lambs (a significant shift in output characteristics) or the creation of a brand for beef.

The second type of strategic decision is to do with the role of the firm in the marketing system. The process of vertical segmentation might lead a firm to the judgment that they do have characteristics of the marketing mix that matter to a level of customers some way through the marketing chain. But they may lack control over output, or achieve a poor flow of returns, which vitiates the prospect of enhancing returns. In this situation, the people running the firm ought to consider ways to increase control or improve the flow of earned premiums.

This could be attempted in various ways. Various forms of vertical coordination or integration may be attempted. Firms may seek to affect public policy in order to change the environment in which the marketing system functions. Historically, the creation of marketing boards and farmer-owned cooperatives have been such attempts.

All such actions have multidimensional effects. A major concern among economists is the move away from the more competitive regimes (in economic terms) that marketing boards, for example, entail. This can lead to farmers, and final customers, being worse off overall. The challenge is to identify innovations to marketing systems that provide net benefits to customers and firms. That is, which strategic responses, involving modifications to marketing systems, make the system more sensitive to customer preferences, while at the same time preserving levels of contestability that ensure appropriate resource allocation by firms within the marketing system?

One approach to vertical coordination to improve sensitivity to consumer preferences is through strategic collaboration. This may take various forms. Strategic alliances involve collaborative behaviour between independent firms at a market level. Joint ventures between independent firms at a market level are a more formalised version of a strategic alliance. Supplier–customer collaboration within marketing systems (sometimes called vertical alliances and including supply chain management) is another form that strategic collaboration may take. In the Australian agribusiness context, there are various examples of strategic alliances. The most thorough analysis available is Hayes et al. (1998) in which strategic alliances in the red meat industry are reviewed, together with consideration of the balance between enhanced market system performance and competitiveness issues.

When firms engage in numerous collaborative relationships, across and within market levels, we see what are called network organisations. This term captures the essence of what occurs in all of these forms of inter-firm collaboration. 'Network organisations' refers to attempts to replicate, to a greater or lesser extent, the flow of market-relevant information that occurs within a single firm that spans similar activities.

The motivation for firms to collaborate is principally to supplant market-based relationships with more informed and customer-oriented integration of productive activity. The objective is to reduce transaction costs and enhance the capacity of the components of the marketing mix to meet customer wants. In systems other than agribusiness, increasing strategic collaboration has been motivated by increasing customer diversity, globalisation and the blurring of market boundaries (for example, as a result of technology convergence in information). These factors are present in agribusiness, of course, but are reinforced as incentives for collaboration by the relatively primitive nature of many agribusiness systems in terms of their market orientation and flows of customer-relevant information to inform innovation.

## MARKETING MANAGEMENT IN AGRIBUSINESS SYSTEMS

The value of the marketing concepts and models we have discussed depends on their usefulness in decision making. Our treatment of marketing in this chapter defines issues that managers of firms, including farms, need to address if they are to make valid marketing management decisions. 'Valid', here, means 'appropriately informed'. Since we have defined all decisions about the marketing mix into 'marketing' management, it is necessary to identify what the criteria are that should apply to decisions about the mix. This includes choice of output and output characteristics within the chosen product category.

Early in the chapter, we noted that marketing is about trying to optimise the marketing mix given the capabilities and competitive situation of the firm. We can now qualify this notion somewhat by adding the insight that to attempt such optimisation of the marketing mix, it is necessary to (a) identify customers, (b) have control over the component characteristics of the mix, and (c) be in a position to make any strategic responses that may be judged appropriate in an attempt to modify the situation.

Taking marketing management one step at a time, then, we can construct a hierarchy of decisions that will be pertinent to any firm in an agribusiness system, including farms.

## Strategic marketing decisions

Strategic marketing decisions span four main areas – what to produce, for whom, when, and with what capabilities – to enable good performance of the output. These areas are intimately connected. Capability to 'deliver the goods' has meaning only in the context of intended output and targeted customers (who define alternative, competing suppliers via their choice criteria). Likewise, choice of what to produce should be informed by the preferences of the customers who are targeted, and by the capacity of the producer to compete successfully in the markets.

The question of what to produce is, for many firms, one with very restricted answers because the skills and resources of the firm are well defined. The skills and resources can be refined, developed or replaced altogether, if necessary, but only over the medium term or longer. Skill and resources are the long-term basis of competitiveness and a meaningful constraint on choice of output. Interestingly, farmers, among firms, commonly have wider choice of output (of 'enterprises' in the farming context) than most small to medium-sized businesses. (The downside of this for farmers is that barriers to entry are low, and supply responses of competitors to price increases correspondingly high, compared to other sectors.)

The choice of customers involves analysis of the kind we have described above. The first aspect of this choice is consideration of just how far down the marketing system it is meaningful to target market levels. The answer to this question is fundamentally binding on a firm. It captures the market reality within which the firm has to operate. If the judgment is that the marketing system could be modified, in time, to make it possible for the firm to aim at market levels closer to the final customer, then this is something the firm might aspire to achieve. For immediate decision-making purposes, however, meaningful target markets are the current reality and must be the basis of choice of target market level(s).

Strategic marketing decisions are a choice. It may be possible and valid for a farm, for example, to aim at retailers rather than the wholesale market. It doesn't have to do so, however. There may be differences in appropriate output characteristics, such as output timing and minimum volumes supplied, that will imply possible changes to production systems and possibly the need for strategic alliances with other farmers. These modifications to farm management may require skills and changes in the nature and sources of production and price risk that don't suit the farmer. Importantly, investment or resources in marketing activity – capital, time, management skill – has to be weighed up like any other investment. It has to be the best use of the limited resources. Simply value or cost adding isn't the objective – profitable value adding is the objective of getting further involved in business activity throughout the marketing chain.

Should the farmer, in this example, choose to continue to focus on the first market level they face in the marketing system, the important implication is that the bases of performance are understood and the target of marketing decisions is clear. The customer is the wholesale buyer. Only the preferences of the wholesale buyers matter when marketing mix decisions are being made. The extent to which these preferences may or may not reflect preferences further down the marketing system is irrelevant to the farmer.

Our analysis of farm management has been on the assumption that the farmer has a clear understanding of what determines the value of output, which we are here describing comprehensively as the marketing mix. This will be so only to the extent that the farmer knows who their customer is – and why: the role of the farmer's output in meeting their needs.

## Agribusiness system marketing developments

There is accelerating change occurring in agribusiness systems in developed economies. It is composed of a variety of ways to inject increased market orientation into the systems, and particularly those providing fresh food. The most significant of such initiatives have retailers using contractual arrangements with farmers and other participants in the system to reduce the volatility in supply volumes, and increase the stability of the quality, of agricultural products. Retailers are well aware of the preferences of final customers and, effectively by offering house-branded fresh food (in the absence of branding), they are in a position to define and benefit financially from improvements in the marketing mix.

Quality assurance schemes of various kinds are employed by retailers linked through the chain to processors and farmers to define their 'consideration sets' with explicit reference to quality dimensions. This makes the retailers a clearly different market segment from retailers and wholesalers who deal in product with more variability of quality. To fresh food producers and processors, such as abattoirs, this implies a choice: they will or will not target this segment. In making this choice, the analytical way of thinking about the questions we have discussed in this text is useful.

A move to 'manage the supply chain' creates quite stark differences between customers of farm output. Retail-level preferences reach right through the marketing system to the farm (the supply chain from a retailer's point of view). However, this doesn't imply that farmers can therefore assume that it is appropriate to target the retail market as customers. Whether that makes sense depends on the answers to the three criteria we presented for vertical segmentation: identifying genuine customers, designing production and optimising production. If retailers seek contracts with farmers, the potential is obviously there for successful targeting of them; whether it is sensible depends on farm enterprise analysis.

Alternatively, clear information may be presented to farmers by customers in the farmer's immediate market about the product characteristics required to tap the benefits of producing the output that the more discriminating retailers within the total retail market require. That is, the retailer preferences cause corresponding segments to exist at the wholesale market level. The appropriateness of aiming to supply the product characteristics demanded is, again, a farm management question.

Related to increasing interest in consumer preferences in agribusiness systems is persistent farmer interest in managing their output as a more precisely targeted 'product', rather than a less targeted 'commodity'. This sounds like an appeal to become more modern and to use the insights that marketing, especially, has to offer. However, the notion begs a great many questions. The analytical frameworks we have discussed lead straight to them.

'Product', being used as the alternative to 'commodity', implies meaningful differentiation and the capturing of the benefits from so doing. Note that the characterisation of outputs as being either commodity or differentiated product is a caricature of reality. All production differs. Products that are characterised as being undifferentiated, and thus commodities, will differ at the very least in place characteristics, with different transport services (costs) associated with the acquisition of the product. Whether further differentiation is plausible and sensible for any product depends on the outcome of vertical segmentation analysis. For many agricultural products, farm-level output is non-differentiable. The basic reason for the inability of farmers to differentiate their product in economically significant ways is sobering: most often, controlled variation in product characteristics of value to customers, other than by location, freight costs, and so on, is not uniquely available to individual farmers. (This is arguably because nature is providing the core production technology.) Meaningful differentiation that is going to attract financial reward will require a new game. This means taking strategic steps to create the possibility of differentiation. Our discussion has, we trust, provided the essential thinking tools that can be employed to judge the plausibility of such an outcome.

## CONCLUSION

Marketing activity starts with production activity. Start with the market is the rule guiding decisions about production. Identify the level of the market, and the customers inhabiting that level, that it is likely to be profitable to supply. Decisions about meeting whatever customer requirements at whatever market level are investment decisions like any other investment decision. The investment of scarce resources – capital, labour, management, time – in extending business activity from one level of the agribusiness system to another level has to be likely to offer a better return than alternative investments for such a move to be justified. The nature of farm product, and the associated marketing systems, often conspires against farmers being able to profitably go very far down the 'marketing' route of identifying customers, controlling the product and optimising the marketing mix.

## CHAPTER SUMMARY

*The effectiveness of marketing is closely tied to the context into which output is delivered. Analysing how the output is viewed by customers, and where the customers are in agribusiness systems, is fundamental to rational decisions related to marketing. It is as possible, and necessary, to be pragmatic and realistic in allocating resources to marketing as to any other aspect of management.*

## REVIEW QUESTIONS

1 Marketing is a specialist activity, just as is production. Marketing requires investment of resources of labour, capital and management. Investment in marketing ought to be appraised just like any other investment in a farm business. Explain.

2 We suggest that marketing isn't as exotic as many people think: it is about defining the relevance of a farm's, or firm's, output to buyers who might become customers and adapting the output to meet their active preferences. Much of the usefulness of reflecting on markets and marketing comes from the multidimensional specification of output and, therefore, preferences. Develop a description of the marketing mix for a food or fibre product with which you are familiar. Choose a market level (such as the one closest to the producing enterprise you have in mind) and identify which customer preferences are most important.

3 Using the information you have from the previous question, analyse the potential for the producing firm or farm to target customers at more distant market levels (for example, retailers or final customers) in the marketing system. That is, could the firm or farm successfully brand its output? If so, which are the specific credence and/or experience characteristics involved?

4 Consider the processes that a particular farm output goes through as it passes through the agribusiness system to the final customer. Taking account of final customer preferences and the kinds of product characteristics involved (experience, and so on), which firms in the system are most likely to wish to coordinate the agribusiness system to improve customer satisfaction? Explain why, and the implications of this for farmers' marketing decisions.

## FURTHER READING

Any recent principles of marketing, introductory marketing or marketing management text will include a thorough discussion of the detailed components of the marketing mix.

Caswell, J.A. and Modjuzska, E.M. 1996, 'Using informational labeling to influence the market for quality in food products', *American Journal of Agricultural Economics*, vol. 78, pp. 1248–53.

Wright, V. 1996, 'Marketing systems, performance and impediments to product differentiation', *Review of Marketing and Agricultural Economics*, vol. 64, no. 2, pp. 135–41.

# 7 Conclusion

Is the farming game a hard way to make a living? Is farming 'a rough shed'? In farming, do farmers always have to travel a hard road to reach pastures of plenty? Do farmers have to be very good at farming to make a go of it? Compared with many other business and professional activities, the answer to these questions is unambiguously 'yes', for all the reasons we have explained to do with biology, technology and scientific development, the arrival of new knowledge, climatic variability, keen competition from other farmers from all around the world, the nature of consumer demand and agricultural commodity markets, and risk and uncertainty. Is the farming game fascinating? Intellectually challenging? Rewarding financially and professionally? Personally fulfilling? Again, the answer is unambiguously 'yes' to these questions, for all the same reasons.

Variously throughout this book we have referred to 'rules' of the farming game. These rules are the product of biological, economic and social phenomena that, taken together, determine the limits of what farmers can do and how they do it. Like in a sporting contest, the rules of farming are never unambiguous, nor are they static. The game is dynamic; so, too, must be the rules. Given below is a list of the most important concepts, ideas, principles, ways of thinking and rules to do with the farming game that, well understood, will benefit anyone having any form of business involvement in farming. At the very least, a sound understanding of these ideas and principles will help people in farming to achieve more of the potential of themselves and their business. A better understanding of these important principles about farming might help some farmers to avoid the main avoidable but potentially business- and career-ending calamities that regularly beset farming businesses. In other cases, a sound understanding of the key principles and ideas of this book will help farmers to recognise where a mismatch may exist between their own situation, the resources they control, and their aspirations and the dictates of the game they are in – and what they might best do about it.

## CONCEPTS, IDEAS, PRINCIPLES, WAYS OF THINKING AND RULES THAT CAN HELP IN UNDERSTANDING THE FARMING BUSINESS

- Economics is the core discipline of farm management analysis. It is the integrating way of thinking.
- The economic approach is the whole farm approach.
- Start with the people.
- Master the technology.
- There is a trade-off between risk and return.
- Above-average returns are only achievable if a business takes above-average risk.
- The status quo is never an option in dynamic farming.
- Marginal thinking, not average thinking, is the key to decisions about resource use: extra cost, extra return, addition to total profit.
- Profit and cash are different – businesses have to be both profitable and liquid.
- Both economic and financial analyses are required, always.
- Growth in net worth is a major objective: the balance sheet is integral to farm management analysis and a good place to start.
- Analysis of a single activity in multi-activity farming can miss much of the story.
- It is better to be roughly right about the whole situation than precisely right about a part of the whole situation.

- In a business world like farming, characterised by great risk and uncertainty, more and more elaborate analyses of decisions add less and less valuable information to the decision.
- Make the decision and make it work, as the world changes both because of and despite the decision.
- Understanding the people and the technology is a necessary, but not sufficient, condition for success – economic understanding is needed.
- Average cost of production for an established business is not a sensible guide to action; marginal cost compared with marginal return is the criterion for action.
- The principle of increasing risk works relentlessly on the balance that is appropriate between gearing and rate of growth over a run of years.
- Businesses succeed where owners are passionate about the activity, when what the owners do is something they can be best at, and where the business can be grown.
- Getting a business to the size where the owner can, and is willing to, focus on the whole business and to forgo the pleasures of the day-to-day tasks is a significant milestone on the road to growth – that is, working *on* the business more than *in* the business.
- The appropriate focus of a farmer, looking through the marketing chain, is on the customer, at whatever level, in the system whose requirements the farmer can genuinely meet.
- Investment of human, physical and financial capital in marketing activity has to be appraised in the same way as any other investment – it has to be the best use of those scarce resources.
- Sound economic analysis and decision analysis can only be built on a sound technical foundation.
- An average is an artificial construct. Distributions of events are what happen. Probabilities of events occurring from a distribution of possible events is the way to think about future happenings.
- Measure what can be measured; think hard about, and spell out clearly, what cannot be measured: these are the guidelines for decision analysis.
- When information is missing that is needed for analysis, turn the question around and define the critical level the unknown factor would have to be for the idea to be worth doing. Then, from what is known, form a judgment about the likelihood of the factor achieving this critical level.

- The future is a different world, so the past is a poor predictor of the future.
- To not act to shift risk is to speculate against future reality.
- Reducing risk reduces the chance of above-average returns and losses.
- Intensification increases the mean and variance of net returns.
- Each farm system is unique; there are as many systems as there are farmers.
- At any time in the farm economy, many different types of farm systems are in existence and, with the right management, any of the wide range of these systems can be profitable. Management of systems is the key, not the nature of the systems as such.
- The economic way of thinking is about benefits and costs, regardless of whether these benefits and costs can all be given a monetary value. Dollars are a convenient unit to use to measure and then compare benefits and costs, but there are always a number of important factors in a decision that cannot convincingly be given a dollar value. Nevertheless, these things are part of total benefits and costs. Farm management analysis is farm benefit-cost analysis.
- Ultimately, you can only really compare yourself with yourself. Comparative analysis and benchmarking the performance of one unique farm business against another farm business that is different and unique is an exercise in futility. Benchmarking cannot answer the question, for any farm business, 'Now what do we do?' Proper farm management analysis answers this question.
- Technical ratios have no economic content. Any technical ratio of input to output in any system can be the most profitable.
- Maximum physical output from a set of resources is never the most profitable amount of output to produce.
- It is better to have an approximate answer to the right question than a precise answer to the wrong question.
- Identifying the question correctly is the essence of good farm management analysis; once the true nature of the problem and the true objectives of the farmer are identified (the right question), the solution choices are usually few and obvious.
- An adviser to a farmer or farm family can never know more than the farmer or farm family about what is good or best for the farmer or farm family.

There is no ending to lists of sensible observations about farming. Common-sense observations (that is, economic thinking) covering essentially similar ground can be found in writings of scholars with a passionate interest in agriculture and the economy going back to at least 4000 BC. As is the nature of things, some of these common-sense principles have been violated forever, too, and continue to be violated to this day. Then, as now, within any population of farmers and other agriculturalists such as researchers, advisers, consultants and people running agricultural-related businesses, there is always a range of competencies and different degrees of economic literacy. As ever, the winners in these games will be those whose thinking passes the tests of common sense and are thus best equipped to play the game.

## THE POWER OF ECONOMIC THINKING*

One manifestation of economic illiteracy is the failure to understand that economics is the core discipline for analysing questions about farm management. Failure to understand that economics is the core discipline of farm management analysis leads public agencies, and others, to inadequate analysis of important questions about farm management. It leads to wrong-headed science equating the maximum with the optimum. It leads to backward-looking accounting misrepresenting the benefits, cost, net benefits and efficiency of farm businesses. It leads to blinkered partial analyses solving small problems while overlooking big ones. With economics as the core discipline of farm management analysis, the awesome analytical power of the whole farm approach allied to marginal thinking is brought to bear on the most important questions of farm management. That is, faced with the challenges of risk and uncertainty and the imperative of imagining the future, farmers have to answer the big questions about timing, choosing new technology, pursuing growth and managing risk.

The great power of economic thinking is, through conceptualisation and abstraction, making sense of resource allocation questions in farm systems characterised by much complexity and powerful dynamics. The logic and rigour of economic thinking act as an antidote to the merely intuitive. If non-economist analysts cannot make sense of the simplified models of farm reality that form the disciplinary-integrating basis of economic analysis, what hope have they of making sense of the vastly more complex real phenomenon? It would be prudent not to be sanguine about cycles of disciplinary fashion and disciplinary momentum, where ideas and methods develop and prevail for a time, before their influence

* The following section draws on B. Malcolm, 2004, 'Where's the Economics? The Core Discipline of Farm Management Has Gone Missing', *Australian Journal of Agricultural and Resource Economics*.

and application decline. When this happens, it's not always because the ideas have been replaced by better ideas and methods.

The challenge for those who continue to work in farm management economics is to re-establish theoretically sound farm management analysis, or farm benefit-cost analysis, based on economics as the core discipline. Education, vigorous vigilance, rigorous professionalism, and enthusiastic and influential collaboration with non-economic disciplinary specialists are thus the most important professional tasks for farm management economists.

For farmers, the challenge remains to master the game. Fortunately, unlike many other business situations, farming doesn't involve direct, head-to-head, conscious rivalry with competing farms. So, for most, there is no competitor fundamentally redefining the game. The rules of the game are relatively stable, even if the conditions under which it is played are volatile. Understanding this, and understanding the nature of the farm business and its links to the environment in which it operates, maximises the chances of farming successfully.

# Appendix: Discounting procedures and tables

In analysing the merit of farm management options, there can be discounting, compounding, sinking funds and annuities to consider. These are further explained in Chisholm and Dillon (1971). The derivation and use of discount tables is fully explained in their work. The diagram in Figure A1.1 captures the essentials of the derivation and use of discounting, compounding and annuities.

Discount tables are based on the sum invested being $1. Figures in the tables show the proportion or multiple of $1 which results for a given combination of number of years and for a particular interest rate.

## Compounding

For example, what will $10,000 grow to if it were invested for 20 years at 5% compound interest? From Table A, 'Growth at compound interest', that is

$$\$10{,}000 \times 2.6533 = \$26{,}533.$$

## Discounting

What will $26,533, received in 20 years' time, be equivalent to in today's dollars, at a discount rate of 5% interest? From Table B, 'Present value of a future lump sum', that is

$$\$26{,}533 \times 0.3769 = \$10{,}000.$$

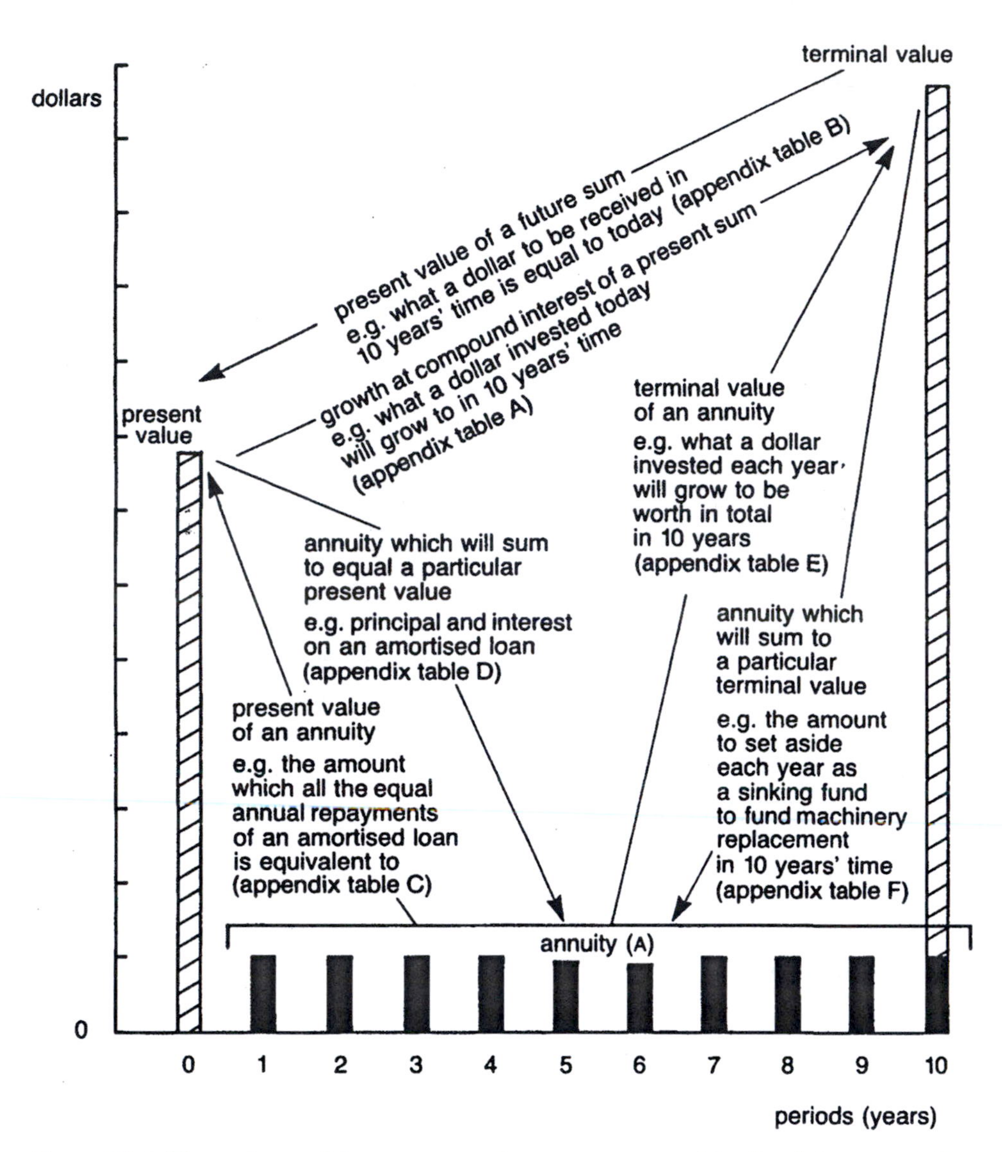


**Figure A1.1** The values of lump sums of money now and in the future, and the present and future equivalent values of sums received or paid in equal instalments

## Annuities

An annuity is an equal sum of money received or spent over a period. If you were to get $1 each year for 20 years, and the relevant interest rate were 5%, then the value (in today's dollars) of this flow of $1 is $12.46. This comes from Table C,

'Present value of an annuity', and equals $1 × 12.4622. Thus, you are equally well-off financially having $12.46 now, or $1 a year for the next 20 years.

Table D shows the 'Annuity whose present value is one'. This is most useful, especially when considering loans. It allows you to work out how much annual interest and principal you have to pay, in equal terms, to amortise (finish or kill) the loan. Thus, for a 20-year loan at 5%, the annuity or amortisation factor is 0.0802. So, $0.08 per year for 20 years will pay off a loan of $1 with interest. If you borrowed $8,000, then your annual repayments would be $8,000 × 0.0802 = $641 p.a. This annuity is also useful for estimating some of the annual costs of owning a machine.

The 'Terminal value of an annuity', in Table E, is the value to which $1 received each year will grow if it and the interest from it are re-invested each year. One dollar per year at 5% for 20 years becomes $33.06. The annuity factor is 33.06.

The 'Annuity whose terminal value is one', from Table F, is also called a 'sinking fund'. It shows the parts of $1 which need to be received each year in order to grow and be $1 at the end of the period. For example, to have $1 in 20 years' time, at a 5% interest rate, then you have to invest $0.03 p.a. This $0.03 p.a. will compound to $1 in 20 years' time. The annuity factor is 0.0302.

## FURTHER READING

Pannell, D.J. 2005, 'Avoiding simplistic assumptions in discounting cash flows for private decisions', in D. Pannell and S. Schilizzi (eds), *Discounting and Discount Rates in Theory and Practice*, Edward Elgar.

**Table A**: Growth at compound interest

Terminal value of a unit of original principal for a term of $n$ periods at a compound rate of interest $i$ per period $(1 + i)^n$

| period | interest rate $i$ | | | | |
|---|---|---|---|---|---|
| $n$ | .01 (1%) | .02 (2%) | .03 (3%) | .04 (4%) | .05 (5%) |
| 1 | 1.0100 | 1.0200 | 1.0300 | 1.0400 | 1.0500 |
| 2 | 1.0201 | 1.0404 | 1.0609 | 1.0816 | 1.1025 |
| 3 | 1.0303 | 1.0612 | 1.0927 | 1.1249 | 1.1576 |
| 4 | 1.0406 | 1.0824 | 1.1255 | 1.1699 | 1.2155 |
| 5 | 1.0510 | 1.1041 | 1.1593 | 1.2167 | 1.2763 |
| 6 | 1.0615 | 1.1262 | 1.1941 | 1.2653 | 1.3401 |
| 7 | 1.0721 | 1.1487 | 1.2299 | 1.3159 | 1.4071 |
| 8 | 1.0829 | 1.1717 | 1.2668 | 1.3686 | 1.4775 |
| 9 | 1.0937 | 1.1951 | 1.3048 | 1.4233 | 1.5513 |
| 10 | 1.1046 | 1.2190 | 1.3439 | 1.4802 | 1.6289 |
| 11 | 1.1157 | 1.2434 | 1.3842 | 1.5395 | 1.7103 |
| 12 | 1.1268 | 1.2682 | 1.4258 | 1.6010 | 1.7959 |
| 13 | 1.1381 | 1.2936 | 1.4685 | 1.6651 | 1.8856 |
| 14 | 1.1495 | 1.3195 | 1.5126 | 1.7317 | 1.9799 |
| 15 | 1.1610 | 1.3459 | 1.5580 | 1.8009 | 2.0798 |
| 16 | 1.1726 | 1.3728 | 1.6047 | 1.8730 | 2.1829 |
| 17 | 1.1843 | 1.4002 | 1.6528 | 1.9479 | 2.2920 |
| 18 | 1.1961 | 1.4282 | 1.7024 | 2.0258 | 2.4066 |
| 19 | 1.2081 | 1.4568 | 1.7535 | 2.1068 | 2.5270 |
| 20 | 1.2202 | 1.4859 | 1.8061 | 2.1911 | 2.6533 |

| $n$ | .06 (6%) | .07 (7%) | .08 (8%) | .09 (9%) |
|---|---|---|---|---|
| 1 | 1.0600 | 1.0700 | 1.0800 | 1.0900 |
| 2 | 1.1236 | 1.1449 | 1.1664 | 1.1881 |
| 3 | 1.1910 | 1.2250 | 1.2597 | 1.2950 |
| 4 | 1.2625 | 1.3108 | 1.3605 | 1.4116 |
| 5 | 1.3382 | 1.4026 | 1.4693 | 1.5386 |
| 6 | 1.4185 | 1.5007 | 1.5869 | 1.6771 |
| 7 | 1.5036 | 1.6058 | 1.7138 | 1.8280 |
| 8 | 1.5938 | 1.7182 | 1.8509 | 1.9926 |
| 9 | 1.6895 | 1.8385 | 1.9990 | 2.1719 |
| 10 | 1.7908 | 1.9672 | 2.1589 | 2.3674 |
| 11 | 1.8983 | 2.1049 | 2.3316 | 2.5804 |
| 12 | 1.9012 | 2.1291 | 2.5182 | 2.8127 |
| 13 | 2.1329 | 2.4098 | 2.7196 | 3.0658 |
| 14 | 2.2609 | 2.5785 | 2.9372 | 3.3417 |
| 15 | 2.3966 | 2.7590 | 3.1722 | 3.6425 |
| 16 | 2.5404 | 2.9522 | 3.4259 | 3.9703 |
| 17 | 2.6928 | 3.1588 | 3.7000 | 4.3276 |
| 18 | 2.8543 | 3.3799 | 3.9960 | 4.7171 |
| 19 | 3.0256 | 3.6165 | 4.3157 | 5.1417 |
| 20 | 3.2071 | 3.8697 | 4.6601 | 5.6044 |

**Table A**: (*cont.*)

| period | interest rate $i$ | | | | | |
|---|---|---|---|---|---|---|
| $n$ | .10 (10%) | .11 (11%) | .12 (12%) | .13 (13%) | .14 (14%) | .15 (15%) |
| 1 | 1.1000 | 1.1100 | 1.1200 | 1.1300 | 1.1400 | 1.1500 |
| 2 | 1.2100 | 1.2321 | 1.2544 | 1.2769 | 1.2996 | 1.3225 |
| 3 | 1.3310 | 1.3676 | 1.4049 | 1.4429 | 1.4815 | 1.5209 |
| 4 | 1.4641 | 1.5181 | 1.5735 | 1.6305 | 1.6890 | 1.7490 |
| 5 | 1.6105 | 1.6851 | 1.7623 | 1.8424 | 1.9254 | 2.0114 |
| 6 | 1.7716 | 1.8704 | 1.9738 | 2.0819 | 2.1950 | 2.3131 |
| 7 | 1.9487 | 2.0762 | 2.2107 | 2.3526 | 2.5023 | 2.6600 |
| 8 | 2.1436 | 2.3045 | 2.4760 | 2.6584 | 2.8526 | 3.0590 |
| 9 | 2.3579 | 2.5580 | 2.7731 | 3.0040 | 3.2519 | 3.5179 |
| 10 | 2.5937 | 2.8394 | 3.1058 | 3.3946 | 3.7072 | 4.0456 |
| 11 | 2.8531 | 3.1518 | 3.4785 | 3.8359 | 4.2262 | 4.6524 |
| 12 | 3.1384 | 3.4984 | 3.8960 | 4.3345 | 4.8179 | 5.3502 |
| 13 | 3.4523 | 3.8833 | 4.3635 | 4.8980 | 5.4924 | 6.1528 |
| 14 | 3.7975 | 4.3104 | 4.8871 | 5.5347 | 6.2613 | 7.0757 |
| 15 | 4.1772 | 4.7846 | 5.4736 | 6.2543 | 7.1379 | 8.1371 |
| 16 | 4.5950 | 5.3109 | 6.1304 | 7.0673 | 8.1372 | 9.3576 |
| 17 | 5.0545 | 5.8951 | 6.8660 | 7.9861 | 9.2765 | 10.7613 |
| 18 | 5.5599 | 6.5435 | 7.6900 | 9.0243 | 10.5752 | 12.3754 |
| 19 | 6.1159 | 7.2633 | 8.6128 | 10.1974 | 12.0557 | 14.2318 |
| 20 | 6.7275 | 8.0623 | 9.6463 | 11.5231 | 13.7435 | 16.3665 |
| $n$ | .16 (16%) | .17 (17%) | .18 (18%) | .19 (19%) | .20 (20%) | .21 (21%) |
| 1 | 1.1600 | 1.1700 | 1.1800 | 1.1900 | 1.2000 | 1.2100 |
| 2 | 1.3456 | 1.3689 | 1.3924 | 1.4161 | 1.4400 | 1.4641 |
| 3 | 1.5609 | 1.6016 | 1.6430 | 1.6852 | 1.7280 | 1.7716 |
| 4 | 1.8106 | 1.8739 | 1.9388 | 2.0053 | 2.0736 | 2.1436 |
| 5 | 2.1003 | 2.1924 | 2.2878 | 2.3863 | 2.4883 | 2.5937 |
| 6 | 2.4364 | 2.5652 | 2.6995 | 3.8398 | 2.9860 | 3.1384 |
| 7 | 2.8262 | 3.0012 | 3.1855 | 3.3793 | 3.5832 | 3.7975 |
| 8 | 3.2784 | 3.5114 | 3.7589 | 4.0214 | 4.2998 | 4.5950 |
| 9 | 3.8030 | 4.1084 | 4.4354 | 4.7854 | 5.1598 | 5.5599 |
| 10 | 4.4114 | 4.8068 | 5.2338 | 5.6947 | 6.1917 | 6.7275 |
| 11 | 5.1173 | 5.6240 | 6.1759 | 6.7767 | 7.4301 | 8.1403 |
| 12 | 5.9360 | 6.5801 | 7.2876 | 8.0642 | 8.9161 | 9.8497 |
| 13 | 6.8858 | 7.6987 | 8.5994 | 9.5964 | 10.6993 | 11.9182 |
| 14 | 7.9875 | 9.0074 | 10.1472 | 11.4198 | 12.8392 | 14.4210 |
| 15 | 9.2655 | 10.5387 | 11.9737 | 13.5895 | 15.4070 | 17.4494 |
| 16 | 10.7480 | 12.3303 | 14.1290 | 16.1715 | 18.4884 | 21.1138 |
| 17 | 12.4677 | 14.4264 | 16.6722 | 19.2441 | 22.1861 | 25.5477 |
| 18 | 14.4625 | 16.8789 | 19.6732 | 22.9005 | 26.6233 | 30.9127 |
| 19 | 16.7765 | 19.7484 | 23.2144 | 27.2516 | 31.9480 | 37.4043 |
| 20 | 19.4608 | 23.1056 | 27.3930 | 32.4294 | 38.3376 | 45.2592 |

**Table B:** Present value of a future lump sum

Present value of a unit amount due in *n* periods at a discount rate of *i* per period

| period | interest rate *i* | | | | |
|---|---|---|---|---|---|
| *n* | .01 (1%) | .02 (2%) | .03 (3%) | .04 (4%) | .05 (5%) |
| 1 | 0.9901 | 0.9804 | 0.9709 | 0.9615 | 0.9524 |
| 2 | 0.9803 | 0.9615 | 0.9426 | 0.9246 | 0.9070 |
| 3 | 0.9706 | 0.9423 | 0.9151 | 0.8890 | 0.8638 |
| 4 | 0.9610 | 0.9238 | 0.8885 | 0.8548 | 0.8227 |
| 5 | 0.9515 | 0.9057 | 0.8626 | 0.8219 | 0.7835 |
| 6 | 0.9420 | 0.8880 | 0.8375 | 0.7903 | 0.7462 |
| 7 | 0.9327 | 0.8706 | 0.8131 | 0.7599 | 0.7107 |
| 8 | 0.9235 | 0.8535 | 0.7894 | 0.7307 | 0.6768 |
| 9 | 0.9143 | 0.8368 | 0.7664 | 0.7026 | 0.6446 |
| 10 | 0.9053 | 0.8203 | 0.7441 | 0.6756 | 0.6139 |
| 11 | 0.8963 | 0.8043 | 0.7224 | 0.6496 | 0.5847 |
| 12 | 0.8874 | 0.7885 | 0.7014 | 0.6246 | 0.5568 |
| 13 | 0.8787 | 0.7730 | 0.6810 | 0.6006 | 0.5303 |
| 14 | 0.8700 | 0.7579 | 0.6611 | 0.5775 | 0.5051 |
| 15 | 0.8613 | 0.7430 | 0.6419 | 0.5553 | 0.4810 |
| 16 | 0.8528 | 0.7284 | 0.6232 | 0.5339 | 0.4581 |
| 17 | 0.8444 | 0.7142 | 0.6050 | 0.5134 | 0.4363 |
| 18 | 0.8360 | 0.7002 | 0.5874 | 0.4936 | 0.4155 |
| 19 | 0.8277 | 0.6864 | 0.5703 | 0.4746 | 0.3957 |
| 20 | 0.8195 | 0.6730 | 0.5537 | 0.4564 | 0.3769 |

| *n* | .06 (6%) | .07 (7%) | .08 (8%) | .09 (9%) |
|---|---|---|---|---|
| 1 | 0.9434 | 0.9346 | 0.9259 | 0.9174 |
| 2 | 0.8900 | 0.8734 | 0.8573 | 0.8417 |
| 3 | 0.8396 | 0.8163 | 0.7938 | 0.7722 |
| 4 | 0.7921 | 0.7629 | 0.7350 | 0.7084 |
| 5 | 0.7473 | 0.7130 | 0.6806 | 0.6499 |
| 6 | 0.7050 | 0.6663 | 0.6302 | 0.5963 |
| 7 | 0.6651 | 0.6227 | 0.5835 | 0.5470 |
| 8 | 0.6274 | 0.5820 | 0.5403 | 0.5019 |
| 9 | 0.5919 | 0.5439 | 0.5002 | 0.4604 |
| 10 | 0.5584 | 0.5083 | 0.4632 | 0.4224 |
| 11 | 0.5268 | 0.4751 | 0.4285 | 0.3875 |
| 12 | 0.4970 | 0.4440 | 0.3971 | 0.3555 |
| 13 | 0.4688 | 0.4150 | 0.3677 | 0.3262 |
| 14 | 0.4423 | 0.3878 | 0.3405 | 0.2992 |
| 15 | 0.4173 | 0.3624 | 0.3192 | 0.2745 |
| 16 | 0.3936 | 0.3387 | 0.2919 | 0.2519 |
| 17 | 0.3714 | 0.3166 | 0.2703 | 0.2311 |
| 18 | 0.3503 | 0.2959 | 0.2502 | 0.2120 |
| 19 | 0.3305 | 0.2765 | 0.2317 | 0.1945 |
| 20 | 0.3118 | 0.2584 | 0.2145 | 0.1784 |

**Table B**: (*cont.*)

| period | interest rate *i* | | | | | | |
|---|---|---|---|---|---|---|---|
| *n* | .10 (10%) | .11 (11%) | .12 (12%) | .13 (13%) | .14 (14%) | .15 (15%) | .16(16%) |
| 1 | 0.9091 | 0.9009 | 0.8929 | 0.8849 | 0.8772 | 0.8696 | 0.8621 |
| 2 | 0.8264 | 0.8116 | 0.7972 | 0.7831 | 0.7695 | 0.7561 | 0.7432 |
| 3 | 0.7513 | 0.7312 | 0.7118 | 0.6930 | 0.6750 | 0.6575 | 0.6407 |
| 4 | 0.6830 | 0.6587 | 0.6355 | 0.6133 | 0.5921 | 0.5717 | 0.5523 |
| 5 | 0.6209 | 0.5934 | 0.5674 | 0.5428 | 0.5194 | 0.4972 | 0.4761 |
| 6 | 0.5645 | 0.5346 | 0.5066 | 0.4803 | 0.4556 | 0.4323 | 0.4104 |
| 7 | 0.5132 | 0.4817 | 0.4523 | 0.4251 | 0.3996 | 0.3759 | 0.3538 |
| 8 | 0.4665 | 0.4339 | 0.4039 | 0.3762 | 0.3506 | 0.3269 | 0.3050 |
| 9 | 0.4241 | 0.3909 | 0.3606 | 0.3329 | 0.3075 | 0.2843 | 0.2629 |
| 10 | 0.3855 | 0.3522 | 0.3220 | 0.2946 | 0.2697 | 0.2472 | 0.2267 |
| 11 | 0.3505 | 0.3173 | 0.2875 | 0.2607 | 0.2366 | 0.2149 | 0.1954 |
| 12 | 0.3186 | 0.2858 | 0.2567 | 0.2307 | 0.2076 | 0.1869 | 0.1685 |
| 13 | 0.2897 | 0.2575 | 0.2292 | 0.2042 | 0.1821 | 0.1625 | 0.1452 |
| 14 | 0.2633 | 0.2320 | 0.2046 | 0.1807 | 0.1597 | 0.1413 | 0.1252 |
| 15 | 0.2394 | 0.2090 | 0.1827 | 0.1599 | 0.1401 | 0.1229 | 0.1079 |
| 16 | 0.2176 | 0.1883 | 0.1631 | 0.1415 | 0.1229 | 0.1069 | 0.0930 |
| 17 | 0.1978 | 0.1696 | 0.1456 | 0.1252 | 0.1078 | 0.0929 | 0.0802 |
| 18 | 0.1799 | 0.1528 | 0.1300 | 0.1108 | 0.0946 | 0.0808 | 0.0691 |
| 19 | 0.1635 | 0.1377 | 0.1161 | 0.0981 | 0.0829 | 0.0703 | 0.0596 |
| 20 | 0.1486 | 0.1240 | 0.1037 | 0.0868 | 0.0728 | 0.0611 | 0.0514 |
| *n* | .17 (17%) | .18 (18%) | .19 (19%) | .20 (20%) | .21 (21%) | .22 (22%) | .23 (23%) |
| 1 | 0.8547 | 0.8475 | 0.8403 | 0.8333 | 0.8264 | 0.8197 | 0.8130 |
| 2 | 0.7305 | 0.7182 | 0.7062 | 0.6944 | 0.6830 | 0.6719 | 0.6610 |
| 3 | 0.6244 | 0.6086 | 0.5934 | 0.5787 | 0.5645 | 0.5507 | 0.5374 |
| 4 | 0.5336 | 0.5158 | 0.4987 | 0.4822 | 0.4665 | 0.4514 | 0.4369 |
| 5 | 0.4561 | 0.4371 | 0.4190 | 0.4019 | 0.3855 | 0.3700 | 0.3552 |
| 6 | 0.3898 | 0.3704 | 0.3521 | 0.3349 | 0.3186 | 0.3033 | 0.2888 |
| 7 | 0.3332 | 0.3139 | 0.2959 | 0.2791 | 0.2633 | 0.2486 | 0.2348 |
| 8 | 0.2848 | 0.2660 | 0.2487 | 0.2326 | 0.2176 | 0.2038 | 0.1909 |
| 9 | 0.2434 | 0.2255 | 0.2090 | 0.1938 | 0.1799 | 0.1670 | 0.1552 |
| 10 | 0.2080 | 0.1911 | 0.1758 | 0.1615 | 0.1486 | 0.1369 | 0.1262 |
| 11 | 0.1778 | 0.1619 | 0.1476 | 0.1346 | 0.1228 | 0.1122 | 0.1026 |
| 12 | 0.1520 | 0.1372 | 0.1240 | 0.1122 | 0.1015 | 0.0920 | 0.0834 |
| 13 | 0.1299 | 0.1163 | 0.1042 | 0.0935 | 0.0839 | 0.0754 | 0.0678 |
| 14 | 0.1110 | 0.0985 | 0.0876 | 0.0779 | 0.0693 | 0.0618 | 0.0551 |
| 15 | 0.0949 | 0.0835 | 0.0736 | 0.0649 | 0.0573 | 0.0506 | 0.0448 |
| 16 | 0.0811 | 0.0708 | 0.0618 | 0.0541 | 0.0474 | 0.0415 | 0.0364 |
| 17 | 0.0693 | 0.0600 | 0.0520 | 0.0451 | 0.0391 | 0.0340 | 0.0296 |
| 18 | 0.0592 | 0.0508 | 0.0437 | 0.0376 | 0.0323 | 0.0279 | 0.0241 |
| 19 | 0.0506 | 0.0431 | 0.0367 | 0.0313 | 0.0267 | 0.0229 | 0.0196 |
| 20 | 0.0433 | 0.0365 | 0.0308 | 0.0261 | 0.0221 | 0.0187 | 0.0159 |

**Table C:** Present value of an annuity

Present value $P$ of an annuity of particular value per period for a term of $n$ periods at a compound interest rate $i$ per period $\left[\frac{(1+i)^n - 1}{i(1+i)^n}\right]$

| period | interest rate $i$ | | | | |
|---|---|---|---|---|---|
| $n$ | .01 (1%) | .02 (2%) | .03 (3%) | .04 (4%) | .05 (5%) |
| 1 | 0.9901 | 0.9804 | 0.9709 | 0.9615 | 0.9524 |
| 2 | 1.9704 | 1.9416 | 1.9135 | 1.8861 | 1.8594 |
| 3 | 2.9410 | 2.8839 | 2.8286 | 2.7751 | 2.7232 |
| 4 | 3.9020 | 3.8077 | 3.7171 | 3.6299 | 3.5459 |
| 5 | 4.8534 | 4.7135 | 4.5797 | 4.4518 | 4.3295 |
| 6 | 5.7955 | 5.6014 | 5.4172 | 5.2421 | 5.0757 |
| 7 | 6.7282 | 6.4720 | 6.2303 | 6.0020 | 5.7864 |
| 8 | 7.6517 | 7.3255 | 7.0197 | 6.7327 | 6.4632 |
| 9 | 8.5660 | 8.1622 | 7.7861 | 7.4353 | 7.1078 |
| 10 | 9.4713 | 8.9826 | 8.5302 | 8.1109 | 7.7217 |
| 11 | 10.3676 | 9.7868 | 9.2526 | 8.7605 | 8.3064 |
| 12 | 11.2551 | 10.5753 | 9.9540 | 9.3851 | 8.8632 |
| 13 | 12.1337 | 11.3484 | 10.6349 | 9.9856 | 9.3936 |
| 14 | 13.0037 | 12.1062 | 11.2961 | 10.5631 | 9.8986 |
| 15 | 13.8650 | 12.8493 | 11.9379 | 11.1184 | 10.3797 |
| 16 | 14.7179 | 13.5777 | 12.5611 | 11.6523 | 10.8378 |
| 17 | 15.5622 | 14.2919 | 13.1661 | 12.1657 | 11.2741 |
| 18 | 16.3983 | 14.9920 | 13.7535 | 12.6593 | 11.6896 |
| 19 | 17.2260 | 15.5785 | 14.3238 | 13.1339 | 12.0853 |
| 20 | 18.0455 | 16.3514 | 14.8775 | 13.5903 | 12.4622 |

| $n$ | .06 (6%) | .07 (7%) | .08 (8%) | .09 (9%) |
|---|---|---|---|---|
| 1 | 0.9434 | 0.9346 | 0.9259 | 0.9174 |
| 2 | 1.8334 | 1.8080 | 1.7833 | 1.7591 |
| 3 | 2.6730 | 2.6243 | 2.5771 | 2.5313 |
| 4 | 3.4651 | 3.3872 | 3.3121 | 3.2397 |
| 5 | 4.2124 | 4.1002 | 3.9927 | 3.8897 |
| 6 | 4.9173 | 4.7665 | 4.6229 | 4.4859 |
| 7 | 5.5824 | 5.3893 | 5.2064 | 5.0329 |
| 8 | 6.2098 | 5.9713 | 5.7466 | 5.5348 |
| 9 | 6.8017 | 6.5152 | 6.2469 | 5.9952 |
| 10 | 7.3601 | 7.0236 | 6.7101 | 6.4177 |
| 11 | 7.8869 | 7.4987 | 7.1390 | 6.8052 |
| 12 | 8.3838 | 7.9427 | 7.5361 | 7.1607 |
| 13 | 8.8527 | 8.3576 | 7.9038 | 7.4869 |
| 14 | 9.2950 | 8.7455 | 8.2442 | 7.7862 |
| 15 | 9.7122 | 9.1079 | 8.5595 | 8.0607 |
| 16 | 10.1059 | 9.4466 | 8.8514 | 8.3126 |
| 17 | 10.4773 | 9.7632 | 9.1216 | 8.5436 |
| 18 | 10.8276 | 10.0591 | 9.3719 | 8.7556 |
| 19 | 11.1581 | 10.3356 | 9.6036 | 8.9501 |
| 20 | 11.4699 | 10.5940 | 9.8181 | 9.1285 |

**Table C**: (*cont.*)

| period | interest rate $i$ | | | | | |
|---|---|---|---|---|---|---|
| $n$ | .10 (10%) | .11 (11%) | .12 (12%) | .13 (13%) | .14 (14%) | .15 (15%) |
| 1 | 0.9091 | 0.9009 | 0.8929 | 0.8849 | 0.8772 | 0.8696 |
| 2 | 1.7355 | 1.7125 | 1.6900 | 1.6681 | 1.6467 | 1.6257 |
| 3 | 2.4868 | 2.4437 | 2.4018 | 2.3611 | 2.3216 | 2.2832 |
| 4 | 3.1699 | 3.1024 | 3.0373 | 2.9745 | 2.9137 | 2.8550 |
| 5 | 3.7908 | 3.6959 | 3.6048 | 3.5172 | 3.4331 | 3.3522 |
| 6 | 4.3553 | 4.2305 | 4.1114 | 3.9975 | 3.8887 | 3.7845 |
| 7 | 4.8684 | 4.7122 | 4.5638 | 4.4226 | 4.2883 | 4.1604 |
| 8 | 5.3349 | 5.1461 | 4.9676 | 4.7988 | 4.6389 | 4.4873 |
| 9 | 5.7590 | 5.5370 | 5.3282 | 5.1317 | 4.9464 | 4.7716 |
| 10 | 6.1446 | 5.8892 | 5.6502 | 5.4262 | 5.2161 | 5.0188 |
| 11 | 6.4951 | 6.2065 | 5.9377 | 5.6869 | 5.4527 | 5.2337 |
| 12 | 6.8137 | 6.4924 | 6.1944 | 5.9176 | 5.6603 | 5.4206 |
| 13 | 7.1034 | 6.7499 | 6.4325 | 6.1218 | 5.8424 | 5.5831 |
| 14 | 7.3667 | 6.9819 | 6.6282 | 6.3025 | 6.0021 | 5.7245 |
| 15 | 7.6061 | 7.1909 | 6.8109 | 6.4624 | 6.1422 | 5.8474 |
| 16 | 7.8237 | 7.3792 | 6.9740 | 6.6039 | 6.2651 | 5.9542 |
| 17 | 8.0215 | 7.5488 | 7.1196 | 6.7291 | 6.3729 | 6.0472 |
| 18 | 8.2014 | 7.7016 | 7.2497 | 6.8399 | 6.4674 | 6.1280 |
| 19 | 8.3649 | 7.8393 | 7.3658 | 6.9380 | 6.5504 | 6.1982 |
| 20 | 8.5136 | 7.9633 | 7.4694 | 7.0247 | 6.6231 | 6.2593 |

| $n$ | .16 (16%) | .17 (17%) | .18 (18%) | .19 (19%) | .20 (20%) |
|---|---|---|---|---|---|
| 1 | 0.8621 | 0.8547 | 0.8475 | 0.8403 | 0.8333 |
| 2 | 1.6052 | 1.5852 | 1.5656 | 1.5465 | 1.5278 |
| 3 | 2.2459 | 2.2096 | 2.1743 | 2.1399 | 2.1065 |
| 4 | 2.7982 | 2.7432 | 2.6901 | 2.6386 | 2.5887 |
| 5 | 3.2743 | 3.1993 | 3.1272 | 3.0576 | 2.9906 |
| 6 | 3.6847 | 3.5892 | 3.4976 | 3.4098 | 3.3255 |
| 7 | 4.0386 | 3.9224 | 3.8115 | 3.7057 | 3.6046 |
| 8 | 4.3436 | 4.2072 | 4.0776 | 3.9544 | 3.8372 |
| 9 | 4.6065 | 4.4506 | 4.3030 | 4.1633 | 4.0310 |
| 10 | 4.8332 | 4.6586 | 4.4941 | 4.3389 | 4.1925 |
| 11 | 5.0286 | 4.8364 | 4.6560 | 4.4865 | 4.3271 |
| 12 | 5.1971 | 4.9884 | 4.7932 | 4.6105 | 4.4392 |
| 13 | 5.3423 | 5.1183 | 4.9095 | 4.7147 | 4.5327 |
| 14 | 5.4675 | 5.2293 | 5.0081 | 4.8023 | 4.6106 |
| 15 | 5.5755 | 5.3242 | 5.0916 | 4.8759 | 4.6755 |
| 16 | 5.6685 | 5.4053 | 5.1623 | 4.9377 | 4.7296 |
| 17 | 5.7487 | 5.4746 | 5.2223 | 4.9897 | 4.7746 |
| 18 | 5.8178 | 5.5338 | 5.2732 | 5.0333 | 4.8122 |
| 19 | 5.8774 | 5.5845 | 5.3162 | 5.0700 | 4.8435 |
| 20 | 5.9288 | 5.6278 | 5.3527 | 5.1009 | 4.8696 |

**Table D:** Annuity whose present value is one

Annuity $A$ whose present value is one for a term of $n$ periods at a compound rate of interest $i$ per period $\left[\frac{i(1+i)^n}{(1+i)^n-1}\right]$

| period | interest rate $i$ | | | | |
|---|---|---|---|---|---|
| $n$ | .01 (1%) | .02 (2%) | .03 (3%) | .04 (4%) | .05 (5%) |
| 1 | 1.0100 | 1.0200 | 1.0300 | 1.0400 | 1.0500 |
| 2 | 0.5075 | 0.5150 | 0.5226 | 0.5302 | 0.5378 |
| 3 | 0.3400 | 0.3468 | 0.3535 | 0.3603 | 0.3672 |
| 4 | 0.2563 | 0.2626 | 0.2690 | 0.2755 | 0.2820 |
| 5 | 0.2060 | 0.2122 | 0.2184 | 0.2246 | 0.2310 |
| 6 | 0.1725 | 0.1785 | 0.1846 | 0.1908 | 0.1970 |
| 7 | 0.1486 | 0.1545 | 0.1605 | 0.1666 | 0.1728 |
| 8 | 0.1307 | 0.1365 | 0.1425 | 0.1485 | 0.1547 |
| 9 | 0.1167 | 0.1225 | 0.1284 | 0.1345 | 0.1407 |
| 10 | 0.1056 | 0.1113 | 0.1172 | 0.1233 | 0.1295 |
| 11 | 0.0965 | 0.1022 | 0.1081 | 0.1141 | 0.1204 |
| 12 | 0.0888 | 0.0946 | 0.1005 | 0.1066 | 0.1128 |
| 13 | 0.0824 | 0.0881 | 0.0940 | 0.1001 | 0.1065 |
| 14 | 0.0769 | 0.0826 | 0.0885 | 0.0947 | 0.1010 |
| 15 | 0.0721 | 0.0778 | 0.0838 | 0.0899 | 0.0963 |
| 16 | 0.0679 | 0.0737 | 0.0796 | 0.0858 | 0.0923 |
| 17 | 0.0643 | 0.0700 | 0.0760 | 0.0822 | 0.0887 |
| 18 | 0.0610 | 0.0667 | 0.0727 | 0.0790 | 0.0855 |
| 19 | 0.0581 | 0.0638 | 0.0698 | 0.0761 | 0.0827 |
| 20 | 0.0554 | 0.0612 | 0.0672 | 0.0736 | 0.0802 |

| $n$ | .06 (6%) | .07 (7%) | .08 (8%) | .09 (9%) |
|---|---|---|---|---|
| 1 | 1.0600 | 1.0700 | 1.0800 | 1.0900 |
| 2 | 0.5454 | 0.5531 | 0.5608 | 0.5685 |
| 3 | 0.3741 | 0.3811 | 0.3880 | 0.3950 |
| 4 | 0.2886 | 0.2952 | 0.3019 | 0.3087 |
| 5 | 0.2374 | 0.2439 | 0.2505 | 0.2571 |
| 6 | 0.2034 | 0.2098 | 0.2163 | 0.2229 |
| 7 | 0.1791 | 0.1856 | 0.1921 | 0.1987 |
| 8 | 0.1610 | 0.1675 | 0.1740 | 0.1807 |
| 9 | 0.1470 | 0.1535 | 0.1601 | 0.1668 |
| 10 | 0.1359 | 0.1424 | 0.1490 | 0.1558 |
| 11 | 0.1268 | 0.1334 | 0.1401 | 0.1469 |
| 12 | 0.1193 | 0.1259 | 0.1327 | 0.1396 |
| 13 | 0.1130 | 0.1197 | 0.1265 | 0.1336 |
| 14 | 0.1076 | 0.1143 | 0.1213 | 0.1284 |
| 15 | 0.1030 | 0.1098 | 0.1168 | 0.1240 |
| 16 | 0.0990 | 0.1059 | 0.1130 | 0.1203 |
| 17 | 0.0954 | 0.1024 | 0.1096 | 0.1170 |
| 18 | 0.0924 | 0.0994 | 0.1067 | 0.1142 |
| 19 | 0.0896 | 0.0968 | 0.1041 | 0.1117 |
| 20 | 0.0872 | 0.0944 | 0.1018 | 0.1095 |

**Table D:** (*cont.*)

| period | interest rate $i$ | | | | | |
|---|---|---|---|---|---|---|
| $n$ | .10 (10%) | .11 (11%) | .12 (12%) | .13 (13%) | .14 (14%) | .15 (15%) |
| 1 | 1.1000 | 1.1100 | 1.1200 | 1.1300 | 0.1400 | 1.1500 |
| 2 | 0.5762 | 0.5839 | 0.5917 | 0.5995 | 0.6073 | 0.6151 |
| 3 | 0.4021 | 0.4092 | 0.4163 | 0.4235 | 0.4307 | 0.4380 |
| 4 | 0.3155 | 0.3223 | 0.3292 | 0.3362 | 0.3432 | 0.3503 |
| 5 | 0.2638 | 0.2706 | 0.2774 | 0.2843 | 0.2913 | 0.2983 |
| 6 | 0.2296 | 0.2364 | 0.2432 | 0.2501 | 0.2572 | 0.2642 |
| 7 | 0.2054 | 0.2122 | 0.2191 | 0.2261 | 0.2332 | 0.2404 |
| 8 | 0.1874 | 0.1943 | 0.2013 | 0.2084 | 0.2156 | 0.2228 |
| 9 | 0.1736 | 0.1806 | 0.1877 | 0.1949 | 0.2022 | 0.2096 |
| 10 | 0.1627 | 0.1698 | 0.1770 | 0.1843 | 0.1917 | 0.1992 |
| 11 | 0.1540 | 0.1611 | 0.1684 | 0.1758 | 0.1834 | 0.1911 |
| 12 | 0.1468 | 0.1540 | 0.1614 | 0.1690 | 0.1767 | 0.1845 |
| 13 | 0.1408 | 0.1481 | 0.1557 | 0.1633 | 0.1712 | 0.1791 |
| 14 | 0.1357 | 0.1432 | 0.1509 | 0.1587 | 0.1666 | 0.1747 |
| 15 | 0.1315 | 0.1391 | 0.1468 | 0.1547 | 0.1628 | 0.1710 |
| 16 | 0.1278 | 0.1355 | 0.1434 | 0.1514 | 0.1596 | 0.1679 |
| 17 | 0.1247 | 0.1325 | 0.1405 | 0.1486 | 0.1569 | 0.1654 |
| 18 | 0.1219 | 0.1298 | 0.1379 | 0.1462 | 0.1546 | 0.1632 |
| 19 | 0.1195 | 0.1276 | 0.1358 | 0.1441 | 0.1527 | 0.1613 |
| 20 | 0.1175 | 0.1256 | 0.1339 | 0.1423 | 0.1510 | 0.1598 |

| $n$ | .16 (16%) | .17 (17%) | .18 (18%) | .19 (19%) | .20 (20%) |
|---|---|---|---|---|---|
| 1 | 1.1600 | 1.1700 | 1.1800 | 1.1900 | 1.2000 |
| 2 | 0.6230 | 0.6308 | 0.6387 | 0.6466 | 0.6545 |
| 3 | 0.4453 | 0.4526 | 0.4599 | 0.4673 | 0.4747 |
| 4 | 0.3574 | 0.3645 | 0.3717 | 0.3790 | 0.3863 |
| 5 | 0.3054 | 0.3126 | 0.3198 | 0.3270 | 0.3344 |
| 6 | 0.2714 | 0.2786 | 0.2859 | 0.2933 | 0.3007 |
| 7 | 0.2476 | 0.2549 | 0.2624 | 0.2698 | 0.2774 |
| 8 | 0.2302 | 0.2377 | 0.2452 | 0.2529 | 0.2606 |
| 9 | 0.2171 | 0.2247 | 0.2324 | 0.2402 | 0.2481 |
| 10 | 0.2069 | 0.2147 | 0.2225 | 0.2305 | 0.2385 |
| 11 | 0.1989 | 0.2068 | 0.2148 | 0.2229 | 0.2311 |
| 12 | 0.1924 | 0.2005 | 0.2086 | 0.2169 | 0.2253 |
| 13 | 0.1872 | 0.1954 | 0.2037 | 0.2121 | 0.2206 |
| 14 | 0.1829 | 0.1912 | 0.1997 | 0.2082 | 0.2169 |
| 15 | 0.1794 | 0.1878 | 0.1964 | 0.2051 | 0.2139 |
| 16 | 0.1764 | 0.1850 | 0.1937 | 0.2025 | 0.2114 |
| 17 | 0.1739 | 0.1827 | 0.1915 | 0.2004 | 0.2094 |
| 18 | 0.1719 | 0.1807 | 0.1896 | 0.1987 | 0.2078 |
| 19 | 0.1701 | 0.1791 | 0.1881 | 0.1972 | 0.2065 |
| 20 | 0.1687 | 0.1777 | 0.1868 | 0.1960 | 0.2054 |

**Table E**: Terminal value of a unit annuity where $\frac{(1+i)^n - 1}{i}$

(Terminal value of an annuity of unit per value per period for a term of $n$ periods at a compound rate of interest $i$ per period)

| period | interest rate $i$ | | | | |
|---|---|---|---|---|---|
| $n$ | .01 (1%) | .02 (2%) | .03 (3%) | .04 (4%) | .05 (5%) |
| 1 | 1.0000 | 1.0000 | 1.0000 | 1.0000 | 1.0000 |
| 2 | 2.0100 | 2.0200 | 2.0300 | 2.0400 | 2.0500 |
| 3 | 3.0301 | 3.0604 | 3.0909 | 3.1216 | 3.1525 |
| 4 | 4.0604 | 4.1216 | 4.1836 | 4.2465 | 4.3101 |
| 5 | 5.1010 | 5.2040 | 5.3092 | 5.4163 | 5.5256 |
| 6 | 6.1520 | 6.3081 | 6.4684 | 6.6330 | 6.8019 |
| 7 | 7.2135 | 7.4343 | 7.6625 | 7.8983 | 8.1420 |
| 8 | 8.2857 | 8.5830 | 8.8923 | 9.2142 | 9.5491 |
| 9 | 9.3685 | 9.7546 | 10.1591 | 10.5828 | 11.0266 |
| 10 | 10.4622 | 10.9497 | 11.4639 | 12.0061 | 12.5779 |
| 11 | 11.5668 | 12.1687 | 12.8078 | 13.4864 | 14.2068 |
| 12 | 12.6825 | 13.4121 | 14.1920 | 15.0258 | 15.9171 |
| 13 | 13.8093 | 14.6803 | 15.6170 | 16.6268 | 17.7130 |
| 14 | 14.9474 | 15.9739 | 17.0863 | 18.2919 | 19.5986 |
| 15 | 16.0969 | 17.2934 | 18.5989 | 20.0236 | 21.5786 |
| 16 | 17.2579 | 18.6393 | 20.1569 | 21.8245 | 23.6575 |
| 17 | 18.4304 | 20.0121 | 21.7616 | 23.6975 | 25.8404 |
| 18 | 19.6147 | 21.4123 | 23.4144 | 25.6454 | 28.1324 |
| 19 | 20.8109 | 22.8406 | 25.1169 | 27.6712 | 30.5390 |
| 20 | 22.0190 | 24.2974 | 26.8704 | 29.7781 | 33.0660 |

| $n$ | .06 (6%) | .07 (7%) | .08 (8%) | .09 (9%) |
|---|---|---|---|---|
| 1 | 1.0000 | 1.0000 | 1.0000 | 1.0000 |
| 2 | 2.0600 | 2.0700 | 2.0800 | 2.0900 |
| 3 | 3.1836 | 3.2149 | 3.2464 | 3.2781 |
| 4 | 4.3746 | 4.4399 | 4.5061 | 4.5731 |
| 5 | 5.6371 | 5.7507 | 5.8666 | 5.9847 |
| 6 | 6.9753 | 7.1533 | 7.3359 | 7.5233 |
| 7 | 8.3898 | 8.6540 | 8.9228 | 9.2004 |
| 8 | 9.8975 | 10.2598 | 10.6366 | 11.0285 |
| 9 | 11.4913 | 11.9780 | 12.4876 | 13.0210 |
| 10 | 13.1808 | 13.8164 | 14.4866 | 15.1929 |
| 11 | 14.9716 | 15.7836 | 16.6455 | 17.5603 |
| 12 | 16.8699 | 17.8885 | 18.9771 | 20.1407 |
| 13 | 18.8821 | 20.1406 | 21.4953 | 22.9534 |
| 14 | 21.0151 | 22.5505 | 24.2149 | 26.0192 |
| 15 | 23.2760 | 25.1290 | 27.1521 | 29.3609 |
| 16 | 25.6725 | 27.8881 | 30.3243 | 33.0034 |
| 17 | 28.2129 | 30.8402 | 33.7502 | 36.9737 |
| 18 | 30.9057 | 33.9990 | 37.4502 | 41.3013 |
| 19 | 33.7600 | 37.3790 | 41.4463 | 46.0185 |
| 20 | 36.7856 | 40.9955 | 45.7620 | 51.1601 |

**Table E:** *(cont.)*

| period | interest rate *i* | | | | | |
|---|---|---|---|---|---|---|
| *n* | .10 (10%) | .11 (11%) | .12 (12%) | .13 (13%) | .14 (14%) | .15 (15%) |
| 1 | 1.0000 | 1.0000 | 1.0000 | 1.0000 | 1.0000 | 1.0000 |
| 2 | 2.1000 | 2.1100 | 2.1200 | 2.1300 | 2.1400 | 2.1500 |
| 3 | 3.3100 | 3.3421 | 3.3744 | 3.4069 | 3.4396 | 3.4725 |
| 4 | 4.6410 | 4.7097 | 4.7793 | 4.8498 | 4.9211 | 4.9934 |
| 5 | 6.1051 | 6.2278 | 6.3528 | 6.4803 | 6.6101 | 6.7424 |
| 6 | 7.7156 | 7.9129 | 8.1152 | 8.3227 | 8.5355 | 8.7537 |
| 7 | 9.4872 | 9.7833 | 10.0890 | 10.4047 | 10.7305 | 11.0668 |
| 8 | 11.4359 | 11.8594 | 12.2997 | 12.7573 | 13.2328 | 13.7268 |
| 9 | 13.5795 | 14.1640 | 14.7756 | 15.4157 | 16.0853 | 16.7858 |
| 10 | 15.9374 | 16.7220 | 17.5487 | 18.4197 | 19.3373 | 20.3037 |
| 11 | 18.5312 | 19.5614 | 20.6546 | 21.8143 | 23.0445 | 24.3493 |
| 12 | 21.3843 | 22.7132 | 24.1331 | 25.6502 | 27.2707 | 29.0017 |
| 13 | 24.5227 | 26.2116 | 28.0291 | 29.9847 | 32.0886 | 34.3519 |
| 14 | 27.9750 | 30.0949 | 32.3926 | 34.8827 | 37.5811 | 40.5047 |
| 15 | 31.7725 | 34.4054 | 37.2797 | 40.4175 | 43.8424 | 47.5804 |
| 16 | 35.9497 | 39.1899 | 42.7533 | 46.6717 | 50.9803 | 55.7715 |
| 17 | 40.5447 | 44.5008 | 48.8837 | 53.7391 | 59.1176 | 65.0751 |
| 18 | 45.5992 | 50.3959 | 55.7497 | 61.7251 | 68.3941 | 75.8363 |
| 19 | 51.1591 | 56.9395 | 63.4397 | 70.7494 | 78.9692 | 88.2118 |
| 20 | 57.2750 | 64.2028 | 72.0524 | 80.9468 | 91.0249 | 102.4436 |

| *n* | .16 (16%) | .17 (17%) | .18 (18%) | .19 (19%) | .20 (20%) |
|---|---|---|---|---|---|
| 1 | 1.1000 | 1.0000 | 1.0000 | 1.0000 | 1.0000 |
| 2 | 2.1600 | 2.1700 | 2.1800 | 2.1900 | 2.2000 |
| 3 | 3.5056 | 3.5389 | 3.5724 | 3.6061 | 3.6400 |
| 4 | 5.0665 | 5.1405 | 5.2154 | 5.2913 | 5.3680 |
| 5 | 6.8771 | 7.0144 | 7.1542 | 7.2966 | 7.4416 |
| 6 | 8.9775 | 9.2068 | 9.4420 | 9.6829 | 9.9299 |
| 7 | 11.4139 | 11.7720 | 12.1415 | 12.5227 | 12.9159 |
| 8 | 14.2401 | 14.7732 | 15.3270 | 15.9020 | 16.4991 |
| 9 | 17.5185 | 18.2847 | 19.0858 | 19.9234 | 20.7989 |
| 10 | 21.3215 | 22.3931 | 23.5213 | 24.7089 | 25.9587 |
| 11 | 25.7329 | 27.1999 | 28.7551 | 30.4035 | 32.1504 |
| 12 | 30.8502 | 32.8239 | 34.9311 | 37.1802 | 39.5805 |
| 13 | 36.7862 | 39.4040 | 42.2187 | 45.2445 | 48.4966 |
| 14 | 43.6720 | 47.1027 | 50.8180 | 54.8409 | 59.1959 |
| 15 | 51.6595 | 56.1101 | 60.9653 | 66.2607 | 72.0351 |
| 16 | 60.9250 | 66.6488 | 72.9390 | 79.8502 | 87.4421 |
| 17 | 71.6730 | 78.9791 | 87.0680 | 96.0217 | 105.9305 |
| 18 | 84.1407 | 93.4056 | 103.7403 | 115.2659 | 128.1167 |
| 19 | 98.6032 | 110.2845 | 123.4135 | 138.1664 | 154.7400 |
| 20 | 115.3797 | 130.0329 | 146.6280 | 165.4180 | 186.6880 |

**Table F**: Annuity whose terminal value is one

Amount $A$ which would need to be set aside each year to yield a particular terminal value at a compound interest rate $i$ per period $\left[\frac{i}{(1+i)^n - 1}\right]$

| period | interest rate $i$ | | | | |
|---|---|---|---|---|---|
| $n$ | .01 (1%) | .02 (2%) | .03 (3%) | .04 (4%) | .05 (5%) |
| 1 | 1.0000 | 1.0000 | 1.0000 | 1.0000 | 1.0000 |
| 2 | 0.4975 | 0.4950 | 0.4926 | 0.4902 | 0.4878 |
| 3 | 0.3300 | 0.3267 | 0.3235 | 0.3203 | 0.3172 |
| 4 | 0.2463 | 0.2426 | 0.2390 | 0.2355 | 0.2320 |
| 5 | 0.1960 | 0.1921 | 0.1883 | 0.1846 | 0.1810 |
| 6 | 0.1625 | 0.1585 | 0.1546 | 0.1508 | 0.1470 |
| 7 | 0.1386 | 0.1345 | 0.1305 | 0.1266 | 0.1227 |
| 8 | 0.1201 | 0.1165 | 0.1124 | 0.1085 | 0.1047 |
| 9 | 0.1067 | 0.1025 | 0.0984 | 0.0945 | 0.0907 |
| 10 | 0.0956 | 0.0913 | 0.0872 | 0.0839 | 0.0795 |
| 11 | 0.0864 | 0.0822 | 0.0781 | 0.0741 | 0.0704 |
| 12 | 0.0788 | 0.0745 | 0.0705 | 0.0665 | 0.0628 |
| 13 | 0.0724 | 0.0681 | 0.0640 | 0.0601 | 0.0564 |
| 14 | 0.0669 | 0.0626 | 0.0585 | 0.0547 | 0.0510 |
| 15 | 0.0621 | 0.0578 | 0.0538 | 0.0499 | 0.0463 |
| 16 | 0.0579 | 0.0536 | 0.0496 | 0.0458 | 0.0427 |
| 17 | 0.0543 | 0.0500 | 0.0459 | 0.0422 | 0.0387 |
| 18 | 0.0510 | 0.0467 | 0.0427 | 0.0389 | 0.0355 |
| 19 | 0.0480 | 0.0438 | 0.0398 | 0.0361 | 0.0327 |
| 20 | 0.0454 | 0.0411 | 0.0372 | 0.0336 | 0.0302 |

| $n$ | .06 (6%) | .07 (7%) | .08 (8%) | .09 (9%) |
|---|---|---|---|---|
| 1 | 1.0000 | 1.0000 | 1.0000 | 1.0000 |
| 2 | 0.4854 | 0.4831 | 0.4808 | 0.4785 |
| 3 | 0.3141 | 0.3110 | 0.3080 | 0.3050 |
| 4 | 0.2285 | 0.2252 | 0.2219 | 0.2187 |
| 5 | 0.1774 | 0.1739 | 0.1705 | 0.1671 |
| 6 | 0.1434 | 0.1398 | 0.1363 | 0.1329 |
| 7 | 0.1191 | 0.1155 | 0.1121 | 0.1087 |
| 8 | 0.1010 | 0.0980 | 0.0940 | 0.0907 |
| 9 | 0.0877 | 0.0840 | 0.0801 | 0.0768 |
| 10 | 0.0763 | 0.0725 | 0.0690 | 0.0658 |
| 11 | 0.0668 | 0.0634 | 0.0601 | 0.0569 |
| 12 | 0.0593 | 0.0559 | 0.0527 | 0.0496 |
| 13 | 0.0530 | 0.0496 | 0.0465 | 0.0436 |
| 14 | 0.0476 | 0.0443 | 0.0413 | 0.0384 |
| 15 | 0.0430 | 0.0398 | 0.0368 | 0.0340 |
| 16 | 0.0390 | 0.0358 | 0.0330 | 0.0303 |
| 17 | 0.0354 | 0.0324 | 0.0296 | 0.0270 |
| 18 | 0.0324 | 0.0294 | 0.0267 | 0.0242 |
| 19 | 0.0296 | 0.0267 | 0.0241 | 0.0217 |
| 20 | 0.0272 | 0.0244 | 0.0218 | 0.0195 |

**Table F:** (*cont.*)

| period | interest rate $i$ | | | | | |
|---|---|---|---|---|---|---|
| $n$ | .10 (10%) | .11 (11%) | .12 (12%) | .13 (13%) | .14 (14%) | .15 (15%) |
| 1 | 1.0000 | 1.0000 | 1.0000 | 1.0000 | 1.0000 | 1.0000 |
| 2 | 0.4762 | 0.4739 | 0.4717 | 0.4695 | 0.4673 | 0.4651 |
| 3 | 0.3021 | 0.2992 | 0.2963 | 0.2935 | 0.2907 | 0.2880 |
| 4 | 0.2155 | 0.2123 | 0.2092 | 0.2062 | 0.2032 | 0.2003 |
| 5 | 0.1638 | 0.1606 | 0.1574 | 0.1543 | 0.1513 | 0.1483 |
| 6 | 0.1296 | 0.1264 | 0.1232 | 0.1201 | 0.1172 | 0.1142 |
| 7 | 0.1054 | 0.1022 | 0.0991 | 0.0961 | 0.0932 | 0.0904 |
| 8 | 0.0874 | 0.0843 | 0.0813 | 0.0784 | 0.0756 | 0.0728 |
| 9 | 0.0736 | 0.0706 | 0.0677 | 0.0649 | 0.0622 | 0.0596 |
| 10 | 0.0627 | 0.0598 | 0.0570 | 0.0543 | 0.0517 | 0.0492 |
| 11 | 0.0540 | 0.0511 | 0.0484 | 0.0458 | 0.0434 | 0.0411 |
| 12 | 0.0468 | 0.0440 | 0.0414 | 0.0390 | 0.0367 | 0.0345 |
| 13 | 0.0408 | 0.0381 | 0.0357 | 0.0333 | 0.0312 | 0.0291 |
| 14 | 0.0357 | 0.0332 | 0.0309 | 0.0287 | 0.0266 | 0.0247 |
| 15 | 0.0315 | 0.0291 | 0.0268 | 0.0247 | 0.0228 | 0.0210 |
| 16 | 0.0278 | 0.0255 | 0.0234 | 0.0214 | 0.0196 | 0.0179 |
| 17 | 0.0247 | 0.0225 | 0.0205 | 0.0186 | 0.0169 | 0.0154 |
| 18 | 0.0219 | 0.0198 | 0.0179 | 0.0162 | 0.0146 | 0.0132 |
| 19 | 0.0195 | 0.0176 | 0.0158 | 0.0141 | 0.0127 | 0.0113 |
| 20 | 0.0175 | 0.0156 | 0.0139 | 0.0123 | 0.0110 | 0.0098 |

| $n$ | .16 (16%) | .17 (17%) | .18 (18%) | .19 (19%) | .20 (20%) |
|---|---|---|---|---|---|
| 1 | 1.0000 | 1.0000 | 1.0000 | 1.0000 | 1.0000 |
| 2 | 0.4630 | 0.4608 | 0.4587 | 0.4566 | 0.4545 |
| 3 | 0.2853 | 0.2826 | 0.2799 | 0.2773 | 0.2747 |
| 4 | 0.1974 | 0.1945 | 0.1917 | 0.1890 | 0.1863 |
| 5 | 0.1454 | 0.1426 | 0.1398 | 0.1370 | 0.1344 |
| 6 | 0.1114 | 0.1086 | 0.1059 | 0.1033 | 0.1007 |
| 7 | 0.0876 | 0.0849 | 0.0824 | 0.0798 | 0.0774 |
| 8 | 0.0702 | 0.0677 | 0.0652 | 0.0629 | 0.0606 |
| 9 | 0.0571 | 0.0547 | 0.0524 | 0.0502 | 0.0481 |
| 10 | 0.0469 | 0.0447 | 0.0425 | 0.0405 | 0.0385 |
| 11 | 0.0389 | 0.0368 | 0.0348 | 0.0329 | 0.0311 |
| 12 | 0.0324 | 0.0305 | 0.0286 | 0.0269 | 0.0253 |
| 13 | 0.0272 | 0.0254 | 0.0237 | 0.0221 | 0.0206 |
| 14 | 0.0229 | 0.0212 | 0.0197 | 0 0182 | 0.0169 |
| 15 | 0.0194 | 0.0178 | 0.0164 | 0.0151 | 0.0139 |
| 16 | 0.0164 | 0.0150 | 0.0137 | 0.0125 | 0.0114 |
| 17 | 0.0139 | 0.0127 | 0.0115 | 0.0104 | 0.0094 |
| 18 | 0.0119 | 0.0107 | 0.0096 | 0.0087 | 0.0078 |
| 19 | 0.0101 | 0.0091 | 0.0081 | 0.0072 | 0.0065 |
| 20 | 0.0087 | 0.0077 | 0.0068 | 0.0060 | 0.0054 |

# References and further reading


Anderson, J.R. and Dillon, J.D. 1992, 'Risk analysis in dryland farming systems', FAO Farm Systems Management Series No. 2, Food and Agricultural Organization of the United Nations, Rome.

Arrow, K. 1992, 'I know a hawk from a handsaw,' in M. Szenberg (ed.), *Eminent Economists: Their life and philosophies*, Cambridge University Press, Cambridge.

Bernstein, P.L. 1996, *Against the Gods: The remarkable story of risk*, John Wiley & Sons, New York.

Boehlje, M.D. and Eidman, V.R. 1983, *Farm Management*, John Wiley & Sons, New York.

Boulding, K.E. 1956, 'General systems theory: The skeleton of science', *Management Science*, vol. 2, pp. 197–208.

Boulding, K.E. 1974, 'General systems as a point of view', in L.D. Singell (ed.), *Collected Papers Vol. 4, Toward a General Social Science*, Colorado Associated University Press, Boulder, Colorado.

Brennan, L.E. and McCown, R.L. 2002, 'Back to the future – reinventing farm management economics in farming systems research', paper presented to the 46th Annual Conference of the Australian Agricultural and Resource Economics Society, Canberra.

Brennan, L. and McCown, R. 2003, 'Making farm management research relevant to farm management practice', invited paper presented to the 47th Annual Conference of the Australian Agricultural and Resource Economics Society, Fremantle.

Campbell, K.O. 1944, 'Production cost studies as a field of research in agricultural economics', *Journal of the Australian Institute of Agricultural Science*, vol. 10, pp. 31–7.

Campbell, K.O. 1957, 'Contemporary agricultural economics Australia', *Review of Marketing and Agricultural Economics*, vol. 25, pp. 24–33.

Candler, W. 1962, 'Production economics and problems of animal production', *Proceedings of the New Zealand Society of Animal Production*, vol. 22, pp. 42–158.

Candler, W. and Sargent, D. 1962, 'Farm standards and the theory of production', *Journal of Agricultural Economics*, vol. 60, pp. 152–63.

Carroll 2004, 'Farm performance from a wealth creation perspective', *Agribusiness*, National Australia Bank, Melbourne.

Chisholm, A.H. and Dillon, J.L. 1971, *Discounting and other Interest Rate Procedures in Farm Management*, ABRI, UNE, Armidale, NSW.

Collins, J. 2001, *Good to Great: Why some companies make the leap . . . and others don't*, Random House, Sydney.

Dillon, J. 1965, 'Farm management as an academic discipline in Australia', *Review of Marketing and Agricultural Economics*, vol. 33, pp. 175–89.

Dillon, J.L. 1976, 'The economics of systems research', *Agricultural Systems*, vol. 1, pp. 1–10.

Dillon, J.L. 1979, 'An evaluation of the state of affairs in farm management', *South African Journal of Agricultural Economics*, vol. 1, pp. 7–13.

Dillon, J.L. 1980, 'The definition of farm management', *Journal of Agricultural Economics*, vol. 31, pp. 257–8.

Ferris, A. and Malcolm, B. 1999, 'Sense and non-sense in dairy farm management economics', *Australasian Agribusiness Perspectives*, vol. 2, paper no. 31, www.agrifood.info.

Fitzgerald, T. 1990, *Between Life and Economics*, Boyer Lectures, ABC Enterprises, Sydney.

Gigerenzer, G. 2002, *Reckoning with Risk: Learning to live with uncertainty*, Penguin Books, London.

Gigerenzer, G. and Todd, P.M. 1999, *Simple Heuristics That Make Us Smart*, Oxford University Press, Oxford.

Goldsworthy, K. 2002, 'After the academy', in P. Craven (ed.), *The Best Australian Essays 2002*, Black Inc., Melbourne, pp. 192–203.

Hayes, G., Malcolm, B., Watson, A., O'Keefe, M. and Thatcher, L. 1998, 'Strategic alliances and the red meat industry in Australia', *Australasian Agribusiness Perspectives*, paper no. 12, www.agrifood.info.

Heady, E.O. 1952, *Economics of Agricultural Production and Resource Use*, Prentice Hall, New York.

Hopkin, J.A., Barry, P.J. and Baker, C.B. 1999, *Financial Management in Agriculture*, The Interstate Printers and Publishers Inc., Danville, Illinois.

Just, R. 2003, 'Risk research in agricultural economics: Opportunities and challenges for the next twenty-five years', *Agricultural Systems*, vol. 75, pp. 123–59.

King, D.A and Sinden, J. 1994, 'Price formation in farmland markets', *Land Economics*, vol. 70, pp. 38–52.

Kingwell, R. 2002, 'Issues for farm management in the 21st century: A view from the West', *Australasian Agribusiness Review*, vol. 10, paper no. 6, www.agrifood.info.

McConnell, D.J. and Dillon, J.L. 1997, *Farm Management for Asia: A systems approach*, Food and Agriculture Organization of the United Nations, Rome.

McCown, R.L. 2001, 'Learning to bridge the gap between science-based decision support and the practice of farming: Evolution in paradigms of model-based research and intervention from design to dialogue', *Australian Journal of Agricultural Research*, vol. 52, pp. 549–71.

McCown, R.L. 2002a, 'Changing systems for supporting farmers' decisions: Problems, paradigms and prospects', *Agricultural Systems*, vol. 74, pp. 179–220.

McCown, R.L. 2002b, 'Locating agricultural decision support systems in the troubled past and socio-technical complexity of models for management', *Agricultural Systems*, vol. 74, pp. 11–25.

McCown, R.L., Hochman, Z. and Carberry, P.S. 2002, 'Probing the enigma of the decision support system for farmers: Learning from experience and from theory', *Agricultural Systems*, vol. 74, pp. 1–10.

McGregor, M., Rola-Rubzen, F., Murray-Prior, R., Dymond, J. and Bent, M. 2003, 'Farm management – bugger the roots, where is the future?', paper presented to the 47th Annual Conference of the Australian Agricultural and Resource Economics Society, Fremantle.

Madden, B.J. and Malcolm, B. 1996, 'Deciding on the worth of agricultural land', *Review of Marketing and Agricultural Economics*, vol. 64, pp. 152–6.

Makeham, J.P. 1968, *Farm Management Economics*, Gill Publications, Armidale, NSW.

Makeham, J.P. and Malcolm, L.R. 1981, *The Farming Game*, Gill Publications, Armidale, NSW.

Makeham, J.P. and Malcolm, L.R. 1993, *The Farming Game Now*, Cambridge University Press, Melbourne.

Malcolm, B. 1990, 'Fifty years of farm management in Australia: Survey and review', *Review of Marketing and Agricultural Economics*, vol. 58, pp. 24–55.

Malcolm, B. 2004, 'Farm management analysis: A core discipline, simple sums, sophisticated thinking', *Australian Farm Business Management Journal*, vol. 1, no. 1, pp. 45–56, www.abfmnetwork.orange.usyd.edu.au.

Malcolm, L.R. 1987, 'Farm management research: Watersheds and widdershins', invited paper presented to the 31st Annual Conference of the Australian Agricultural and Resource Economics Society, Adelaide.

Malcolm, L.R. 1990, 'Fifty years of farm management in Australia: Survey and review', *Review of Marketing and Agricultural Economics*, vol. 58, pp. 24–55.

Malcolm, L.R. 2000, 'Farm management economic analysis: A few disciplines, a few perspectives, a few futures', invited paper presented to the 44th Annual Conference of the Australian Agricultural and Resource Economics Society, Sydney.

Martin, S. and Woodford, K. 2003, 'The farm management profession in New Zealand: Where are our roots?', invited paper presented to the 47th Annual Conference of the Australian Agricultural and Resource Economics Society, Fremantle.

Mauldon, R.G. and Schapper, H. 1970, 'Random numbers for farmers', *Journal of the Australian Institute of Agricultural Science*, vol. 36, pp. 279–84.

Mishan, E.J. 1982, *Cost Benefit Analysis: An informal introduction*, George Allen & Unwin, London.

Mullen, J.D. 2002, 'Farm management in the 21st century', *Australasian Agribusiness Review*, vol. 10, paper no. 5, www.agrifood.info.

Murray-Prior, R. and Wright, V. 2001, 'Influence of strategies and heuristics on farmers' response to change under uncertainty', *Australian Journal of Agricultural and Resource Economics*, vol. 45, pp. 573–98.

Musgrave, W. 1976, 'Problems of change in Australian agricultural economics', *Australian Journal of Agricultural Economics*, vol. 20, pp. 133–44.

Pannell, D.J., Malcolm, L.R. and Kingwell, R.S. 2000, 'Are we risking too much? Perspectives on risk in farm modelling', *Agricultural Economics*, vol. 23, pp. 69–78.

Ronan, G. 2002, 'Delving and divining for Australian farm management agenda: 1970–2010', *Australasian Agribusiness Review*, vol. 10, paper no. 7, www.agrifood.info.

Schultz, T.W. 1939, 'Theory of the firm and farm management research', *American Journal of Agricultural Economics*, vol. 21, pp. 570–86.

Shiller, R. 2003, *The New Financial Order: Risk in the 21st century*, Princeton University Press, Princeton.

Trapnell, L.N. 1998, 'An investigation of the net benefits from liming acid soils', Masters Thesis, Institute of Land and Food Resources, University of Melbourne.

Wright, V. 1983, 'Some bounds to the relevance of decision theory', *Australian Journal of Agricultural Economics*, vol. 27, pp. 221–30.

# Glossaries

## FARM MANAGEMENT ECONOMICS

Listed below are some of the more common, general terms used in farm management economics. Many other, more specific terms are defined fully in the text (terms relating to futures markets are defined on pp. 198–200).

**Activity** A particular method of producing a commodity. A more specific term than 'enterprise' – for example, spring wheat, winter-fattened steers.

**Activity gross income** The total value of the output of a farm activity whether the output is sold or not.

**Activity gross margin** Activity gross income minus the variable costs of that activity.

**Agribusiness** Businesses that are closely related to agricultural production activities.

**Amortised loan** A loan that is repaid in equal instalments of principal and interest, with the interest and principal components of the repayment instalment varying as the loan reduces.

**Annuity** A sum of money received or used every year that is equivalent to a larger sum at the present time or at a future time – for example, the equal annual sum that repays the interest and principal on an amortised loan. An annual sum over a number of years that is equivalent to the net present value of an investment project that runs over the same number of years.

**Break-even analysis** Testing the key elements of a budget for a proposed change to a system to determine the level at which the key elements of the proposal make the overall result just equal to the result from an alternative action.

**Budget** A detailed statement of a future plan of action detailing the expected costs and benefits.

**Budget control** The process of comparing the actual performance of an aspect of farm production against the performance that was expected when the budget was drawn up.

**Business health** The state of and prospects for profit, financial viability and growth of a business.

**Capital** Items that contribute to production over at least a medium-term time period, such as for more than a year – for example, tractor, land, lime fertiliser, structures and equipment.

**Capital gains** Increase in the value of capital items due to a rise in their market value.

**Capital investment** Funds used to acquire assets such as equipment, land or stock, or used on improvements, that have a life of more than one year and add to the productive capacity of the farm.

**Cash flow** The movement of funds in and out of the hands of an enterprise or individual farmers.

**Cash flow budget** A budget of the expected cash in (receipts) and cash out (payments) associated with a particular farm plan.

**Cast for age (CFA)** A reject animal that is past its economic life for particular conditions.

**Comparative analysis** Comparison of the performance of a particular farm with some 'standard' level of performance. (Usually the 'standard' is the average performance of a group of broadly similar farmers.) Benchmarking is a term used to describe a similar approach.

**Compound interest rate** The rate of interest used in compounding.

**Compounding** The way a sum of money grows to a larger sum by adding interest, then reinvesting the larger sum to earn interest again – that is, calculation of the equivalent future value of a present sum.

**Contingency allowance** Allowance included in budgets to cover unexpected costs.

**Core discipline** An integrating body of knowledge. Integrates knowledge from many disciplines into comprehensive understanding, analysis and explanation. The core discipline of economics makes it possible to understand, analyse and explain the operation of a whole farm system.

**Cost-price squeeze** The phenomenon of farmers' real costs for their inputs rising and prices they receive for their products being static, falling, or rising at a slower rate than real costs are rising. It means that farmers have to increase their productivity to remain profitable.

**Debt servicing capacity** Annual whole farm net cash flow available to meet interest and loan repayments.

**Decision analysis** A procedure for rigorously and methodically weighing up the expected benefits and costs of a possible action. A way of ensuring that decision makers make decisions that are consistent with their personal beliefs about the risks they face and their personal preferences for possible consequences from the decision.

**Decision tree** A diagrammatic representation of the alternative and sequential actions of a risky decision problem.

**Demand** The amounts of a product or service that consumers wish to buy at a range of prices.

**Depreciation** The loss in value of capital items as they get used and become older.

**Development budget** A budget of cash flows used to assess expected profitability and financial feasibility when planning major farm system changes that will take some time to reach full capacity.

**Diminishing marginal returns** The phenomenon where increases in variable inputs added to fixed inputs in a production process results in smaller and smaller increases in total output. The principle of diminishing returns indicates that variable input should be added to the production process so long as the extra return exceeds the extra cost, and the maximum total profit is at the point where extra return equals extra cost.

**Discounting** The process of adjusting the value of a benefit or cost to be received in the future to their equivalent value in the present time.

**Discounting factor** The adjustment factor used to adjust future values to present values, given by the formula $1/(1+r)^n$, where '*r*' is the discount rate and '*n*' is the number of the year in the future in which the benefit or cost occurs.

**Economic efficiency** Measured by percentage return on all the capital invested in the business.

**Elasticity of demand** The responsiveness of the quantity of a product or service that people demand to a change in price or a change in income. Price elasticity is measured as a percentage change in quantity demanded divided by percentage change in price. Income elasticity is measured as a percentage change in quantity demanded divided by percentage change in income.

**Enterprise** The production of a particular commodity or group of related commodities. A general term – for example, wheat, beef.

**Equity** The value of assets minus liabilities. Also known as net worth. What the business owes the owners.

**Equity percentage** Farm equity capital as a percentage of total farm capital – that is, (assets minus liabilities/assets) × 100/1.

**Farm benefit cost analysis** The budgeting process of evaluating the benefits and costs and the net benefits of an investment to change a farm system.

**Finance budget** A budget showing the flows of cash in and out, in nominal dollars. Identifies borrowings that are needed, and interest and principal repayments.

**Fixed capital** Land, buildings, bores, irrigation equipment, and so on, that cannot easily be moved.

**Fixed costs** Costs that must be met and are not affected by the amount of output produced in a year. Also called overhead costs. They are unavoidable costs in the short to medium term.

**Futures** Quantities of a commodity of defined quality for delivery at an agreed future date. (Chapter 5, pp. 198–200, contains a glossary of key terms relating to futures markets.)

**Gearing** The ratio of debt to equity. Has implications for debt servicing ability and rate of growth of equity. (Also called leverage.)

**Gross margin** Gross income minus variable costs. Can be whole farm gross margin, as in a whole farm budget, or activity gross margin.

**Gross margins planning** A procedure whereby activities are selected sequentially on the basis of the highest gross margin from a unit of only one key constraint, usually land.

**Growth** Increase in net worth (wealth) over time. Measured as change in equity, or net farm income minus tax and consumption above operator's allowance.

**Hedging** Insuring against a loss on holding stocks of a commodity due to a price change during the period of ownership.

**Income elasticity** The responsiveness of demand to changes in income.

**Inflation** An increase in the supply of money in relation to the supply of goods and services available and, in consequence, a decline in the purchasing power or value of currency.

**Innovation** Changing how a farm system operates.

**Interest** The annual sum a lender charges someone who borrows funds off them, expressed as a percentage of the sum borrowed – for example, 10% p.a. interest on $100,000 borrowed.

**Interest-only loan** A loan where the borrowed capital is not intended to be repaid on a regular and gradual basis over the life of the loan and instead annual interest is paid on the full amount of the borrowed capital for the life of the loan.

**Intermediate activity** The production of a commodity that is not sold directly but becomes an input for other activities of the farm – for example, crop stubble for grazing.

**Internal rate of return** The discount rate at which the present value of future benefits from a project equals the present value of total costs of the project.

**Investment appraisal** An evaluation of the profitability and financial feasibility of a potential investment.

**Linear programming** A mathematical, computer-based, farm planning technique that determines the combination of activities that maximises total gross margin or profit, or minimises costs. There are usually a range of alternative solutions (farm plans) that produce a total gross margin or profit very close to the optimum, and it is not so much what you do as how you do it that is the practical decision rule.

**Liquidity** Cash or near-cash reserves. Relates to the ability of a business to service debt.

**Livestock feed budget** A budget comparing current and predicted feed requirements of livestock with the feed available and expected supply.

**Livestock gross income** The value of livestock production in the form of animals and produce, adjusted for inventory changes.

**Livestock trading schedule** A budget used to estimate the annual contribution to gross income from the trading of animals by sales and purchases, and births and deaths, and changes in the numbers and value of livestock on hand, from opening number and value to closing number and value. Captures the effects of animal depreciation and appreciation, as well as natural increase.

**Machinery replacement allowance** Sum deducted from net cash flow each year so that funds are available to replace capital items when they are worn out.

**Marginal** Economists' term for 'extra' or 'added'. The principle of marginality refers to the profit-maximising level of operation where the marginal revenue from production equals the marginal cost of production.

**Marginal cost** The extra cost added to total cost from using an extra unit of a variable input; or the extra cost incurred in growing or selling an additional unit of product.

**Marginal product** The change in output arising from using an extra unit of a variable input.

**Marginal revenue** The extra net income obtained from selling one additional unit of product.

**Marginal value product** The value of an extra unit of output. The marginal physical product of a unit of output times the price per unit of the product.

**Marketing margin** The difference between the purchase price and resale price of a product between two levels in a marketing chain. Indicates the cost of adding services to products.

**Monopoly** There is only one seller of a product or service.

**Net cash flow** The difference between the money received and the money spent in any one period (week, month, year).

**Net farm income** Operating profit minus interest. Also called net profit. It is the return on the owner's capital.

**Net present value (NPV)** The difference between the present value of all benefits and the present value of all costs of an investment, with the present values of benefits and costs calculated using a particular discount rate.

**Net worth** The value of total assets minus the value of total liabilities (equity).

**Nominal terms** Dollar values or interest rates that include an inflation component.

**Oligopoly** There are only a few sellers of a certain product or service so that each will be affected substantially by a change in policy on the part of another.

**Operating costs** Variable costs plus overhead costs.

**Operating profit** Gross income minus variable and fixed (overhead) costs. It is the return on all the capital invested.

**Opportunity cost** The opportunity cost of a farm management decision is the amount of net benefit that is given up by choosing one alternative action rather than some other action.

**Overhead (fixed) costs** Costs that do not vary as the level of production or mixture of activities changes. They are unavoidable costs in the short to medium term.

**Parameter** Any factor that has an important effect on profit (such as yield, price, land area or direct cost).

**Parametric budget** A planning technique that takes varying prices and yields into account.

**Partial budget** A budget drawn up to estimate the effect on profit of a proposed change affecting only part of the farm. Used to estimate the extra return on extra capital invested.

**Payoff matrix** A table showing the probabilities of and outcomes due to different acts and states of nature occurring.

**Principal** The amount of capital borrowed when a loan is taken out. Principal repayments are the amounts of capital repaid to settle a debt.

**Principle of increasing risk** The more highly geared the business, the more rapidly equity grows when things go well, but equity declines at an even faster rate when things go badly.

**Production function** The relationship between the level of inputs and the level of output for some production process. Also called a response function.

**Real terms** Dollar values or interest rates that have no inflation component.

**Return on total assets** Operating profit expressed as a percentage of the value of total farm assets.

**Return on total capital** The annual operating profit expressed as a percentage of the total capital invested in the business over the year. (Total capital can be capital at the start, end or, better, an average of the start and end capital value.) Percentage return to total capital is the measure of economic efficiency.

**Risk** A situation with uncertain outcomes, but a case where some probabilities can be formed about the outcomes. This is in contrast to uncertainty, where no probabilities can be formed about uncertain events happening.

**Risk premium** An amount that a person requires above a risk-free return before being willing to accept a risk.

**Scenario analysis** A way of imagining a set of combined circumstances in the future and the implications for important decision criteria, such as profit, growth, financial feasibility and risk.

**Sensitivity testing** Checking the effect on a planned outcome of a change in one of the factors (parameters, coefficients) that affects that outcome.

**Solvent** Assets exceed debts.

**Spot price** The price for a product available for immediate delivery.

**Stock equivalents** Units used in livestock feed budgeting, whereby the energy needs of different categories of livestock are expressed in terms of one type of livestock – for example, dry sheep equivalent (DSE).

**Subjective probability** The strength of belief an individual will hold about the chance of a particular event occurring.

**Substitution** The giving up of one enterprise or activity, or input, for another enterprise or activity or input.

**Supply** The amounts of a product or service that will be offered for sale at a range of prices.

**Technical efficiency** A ratio of the quantity of physical output to quantity of physical input. Does not indicate profitability or economic efficiency.

**Term loan** A loan that is to be repaid in equal annual instalments of principal, with interest charged on the reducing outstanding balance of the loan.

**Variable costs (also direct costs)** Costs that change directly according to the amount of output of the activity – for example, fuel and seed.

**Whole farm approach** The farm management economic method. Understanding and analysing the farm system – the human, technical, economic, financial, risk, institutional elements – as a whole system.

**Whole farm cash budget** A budget showing the expected flows of cash in and out of the business for the coming year.

**Whole farm planning** Planning for the whole farm, as distinct from partial budget planning.

**Whole farm profit budget** Budget showing the expected outcomes of a farm plan, in terms of the entire farm's profitability for the coming year.

**Working capital** Capital needed for the day-to-day operation of a farm. Usually funded by relatively short-term borrowings related to the length of the production cycle – for example, bank overdraft facility and bank bills.

## AGRIBUSINESS MARKETING

**Agribusiness system** See *Marketing system*.

**Brand** A promise of the presence of choice-relevant characteristics. A search characteristic itself, it can only add value if customer choice criteria include experience or credence characteristics.

**Choice criteria** The very small number of criteria that a customer uses to make the final choice between products within the consideration set, if there is more than one in the set.

**Consideration set** The group of products or brands called to mind by a consumer when a trigger need prompts them to acquire a product. Satisfaction with a specific brand may cause it to be the only product in the set. Also called the evoked set.

**Credence characteristics** Characteristics of a product the presence of which the customer can never personally verify. Third parties have to be relied upon for such information. See *Search characteristics*.

**Customers** Individuals or firms who purchase a product. They are not necessarily the consumer of the product they purchase. (They may be acting as an agent for consumers.)

**Evoked set** See *Consideration set*.

**Exchange facilitation characteristics** Aspects of output (the marketing mix) that essentially enable successful exchange but are ephemeral since they do not contribute to customer satisfaction from product consumption. Price, place and promotion fall into this category, usually.

**Experience characteristics** Characteristics of a product the presence of which the customer can identify only after consumption. See *Search characteristics* and *Credence characteristics*.

**Market levels** Stages in marketing systems where changes in ownership of a product occur.

**Marketing** All decisions made by an entity that determine the characteristics of the output (the marketing mix) presented to potential customers.

**Marketing channel** The path between a specific organisation and its customers. One segment only of a marketing system.

**Marketing mix** The entire set of characteristics of a product a producer presents to a potential customer. Often defined as being composed of product, price, place and promotion (the four Ps).

**Marketing system** The entire system composed of entities contributing to the production of a category of product. Also, in agricultural product contexts, called agribusiness system.

**Mix control** The precursor to optimisation of the marketing mix is control over the status of each output characteristic. Low control, as is often true of farms, is problematic for the effective application of marketing to an organisation.

**Search characteristics** Characteristics of a product the presence of which the customer can identify prior to purchase and consumption. See *Experience characteristics* and *Credence characteristics*.

**Segmentation** The disaggregation of a market into sub-markets composed of customers with relevant similarities. In this text, segmentation is considered in both horizontal (market-level) and vertical (marketing system) contexts.

**Strategic marketing** Long-lived decisions about what kinds of product to be produced for which customer segments.

**Targeting** The selection of market segments, or market levels, to whose preferences output characteristics (the marketing mix) will be oriented.

**Trigger need** Defined in this text as the need that initiates the seeking of a product by a consumer.

# Index

CPSIA information can be obtained at www.ICGtesting.com
Printed in the USA
BVOW01s2024060714

357941BV00016B/74/P